The Transmission of Epidemic Influenza

The Transmission of Epidemic Influenza

R. Edgar Hope-Simpson, F.R.C.G.P.

Cirencester Epidemiological Research Unit
Cirencester, Gloucestershire, England

Plenum Press • New York and London

Library of Congress Cataloging-in-Publication Data

Hope-Simpson, R. Edgar.
The transmission of epidemic influenza / R. Edgar Hope-Simpson.
p. cm.
Includes bibliographical references and index.
ISBN 0-306-44073-3
1. Influenza--Transmission. 2. Influenza--Epidemiology.
I. Title.
[DNLM: 1. Influenza--transmission. WC 515 H785t]
RA644.I6H66 1992
614.5'18--dc20
DNLM/DLC
for Library of Congress 92-3221
CIP

ISBN 0-306-44073-3

A Division of Plenum Publishing Corporation
233 Spring Street, New York, N.Y. 10013

Printed in the United States of America

To Eleanor, beloved companion

Years, precious years
Full of laughter and tears;
Shared tears that years after
So often change to laughter.

Foreword

THE PLAGUE YEARS

Mankind has always been fascinated by "origins," and biologists are no exception. Darwin is our most famous example. What is the origin of mankind, of species, of infectious diseases? In the last few years we have seen the emergence and spread of some apparently "new" viruses, such as HIV-1 and the virus causing bovine spongiform encephalomyelopathy. But are these, in fact, entirely new agents, or mutated forms of "old" viruses that have evolved along with us for eons? Edgar Hope-Simpson could not have written this book at a more opportune moment. He is a firm believer in gradual evolution, rather than the sudden arrival of new agents. I suspect that he would also have a naturalist's Darwinian approach for the origin of AIDS.

It has been a source of some amazement to me over the years how even the most innovative scientists conform to a current hypothesis. Pioneer thinking comes more easily to persons outside the scientific mainstream. Edgar Hope-Simpson has always struck me as a modern-day naturalist of the classic style, observant and perhaps a little maverick in line of thought. Certainly, the central hypothesis propounded in this book will be controversial to many scientists. From his unique citadel, the Epidemiological Research Unit in Cirencester, he has carefully reexamined mortality data from old records as well as new. While most influenza virologists have rushed onward, on an unstoppable express train, from one brilliant technology to another, from polyacrylamide gels to separate influenza genes to probes that will detect virus in throat washes of an afflicted patient, Hope-Simpson has had more time to think and observe. His book is almost an "origin of species" approach. To my mind the core of his idea is the shrewdly observed phrase that "man is most in contact with man." Then, while others of us restlessly search the species of the world for new influenza A viruses from whales, seals, birds, and ponds, and propound ideas of "new" viruses emerging from our animal friends, he stays at home and ponders alternative ideas.

Is it really likely that the origin of new pandemic strains of flu is a remote gosling flapping along on the migratory route, or a family pig living in the bedroom *cum* kitchen of a Nientisin family, or in a whale plunging in the Pacific? Or would the origin more likely be our neighbors at home? The author discusses two ideas in the book. First, epidemics are caused by a previous "seeding" of virus in the community. Symptomless carriers spread the virus around, which then causes an epidemic to break out. Second, do these great "new" pandemics of flu, which came out of the blue in 1918, 1957, and 1968, recycle from our older compatriots? Both ideas will have proponents and opponents, but these are worthy of scientific airing. The events of the winter of 1990, when in the United Kingdom alone 26,000 persons died of influenza, warn us that the virus still has a deadly punch. For Daniel Defoe, in the Plague Year, only one disease was important; in the era of AIDS we should not forget our other plagues, influenza and even measles and malaria. Millions of children die yearly from these diseases. Infectious disease has *not* been conquered. The plague year is every year, even in our new technological world. But at the same time there is a new naturalist interest, in Amazon rain forests, ecology, pollution, and in a single world. Understanding an old plague can do nothing but help us with an understanding of all diseases. The "Newe acquaintance" of the court of Elizabeth I is still with us in the reign of Elizabeth II. So it is time for new thoughts and a stirring of scientific complacency. This book provides the seed required for crystallization. Some brilliant minds of the world have tried to untangle influenza, from Burnet to Andrewes, from Stuart-Harris to Francis, Pereira, and Laver. Hope-Simpson might be wrong, but I feel that we have in this book something new, a reappraisal from a unique mind. What I would like to see happen now is the application of modern technology, PCR (polymerase chain reaction), and molecular probes to reexamine the origin of the last pandemic virus of humans.

Professor J. S. Oxford
The London Hospital Medical College
London, England

Acknowledgments

The work that is discussed in this book was begun nearly 60 years ago during the devastating influenza epidemic of the winter of 1932–1933, and has continued as opportunity offered ever since. Through the kindness of my publisher I can at last publicly thank the many persons and organizations who have given me encouragement, advice, and much practical help. It is not possible to name them all, and those whose names are omitted for lack of space must understand that their kindness is not forgotten. When persons are named, this must not be taken as an indication that they agree with the speculations in the book.

Patients of my general practice, first in Beaminster, Dorset, from 1932 to 1945, and then those in Cirencester, Gloucestershire, from 1946 to 1976, cooperated in the studies that are reported in several of the following chapters, and our findings would not have been possible without their help.

In 1947, Sir Harold Himsworth and the late Sir Graham Wilson instituted the Cirencester Epidemiological Research Unit as described in Chapter 1 and continued to watch over the development, adding the microbiological laboratory in 1961 and so providing an unusual opportunity to authenticate virologically the observations carried out from general practice. The late Dr. Marguerite Pereira and her colleagues at the PHLS Central Laboratory, Colindale, London, provided detailed identification of influenza virus isolates and much valued advice.

The staff of the general practice and those of the research unit were closely integrated, adding considerably to their heavy labors. Miss Joyce Dawson was responsible for coordinating the observational studies, the secretarial work, and much of the practical collection of data for 28 years. Mrs. Bettie Neal, who succeeded her, has been responsible for successive revisions and for the final manuscript of this book. Miss Dorothy Cutler (Mrs. Cliffe), the first field worker of the unit until her marriage in 1950, was succeeded by Miss Janet Edmonds. The Gloucestershire County Health Authority kindly seconded Health Visitors to assist in the meticulous household studies of the first epidemics of Hong Kong influenza in this area. Some of the findings are discussed in Chapter 7.

The research into parish burial archives and death registers (Chapter 17) was supported by grants from the Wellcome Trust and later by the Department of Health and Social Security. We are grateful to the County Archive departments of Gloucester, Devon, Dyfed, Cumbria, Northumbria, and Norfolk, and to Dr. David Smith, Dr. W. A. Seaman, and Mr. D. G. Sandford. The Registrar General kindly permitted access to death registers and we are particularly grateful for the valuable help of Mr. F. J. Petrie and Mrs. Hansson of the Cirencester Register Office.

Miss Hazel Spurrier and her staff at the Medical Library, Princess Margaret Hospital (PMH), Swindon, have for many years kept us abreast of the vast influenza literature, helped by the *Weekly Summary* from the PHLS Library, Colindale, and the *Monthly Influenza Bibliography* from the library of the National Institute for Medical Research, Mill Hill. Mr. St. John A. Fallows and his staff at the PMH Department of Medical Photography have also helped us for many years, and Miss Kelene Huntley has redrawn all the illustrations for this volume.

I am grateful for help and advice from Professor E. A. Wrigley, Dr. Richard Wall, and others of the Cambridge Group for the History of Population and Social Structure; Dr. A. D. Barbour of the Statistical Laboratory, Cambridge, for hospitality and percipient suggestions; Dr. Alan Welsford, County Librarian, Cirencester; Dr. David Tyrrell; Professor David Miller; Dr. A. S. Beare; the late Sir Christopher Andrewes; the late Dr. John Dingle and his colleagues at Case Western Reserve University, Cleveland, Ohio; Dr. Alexander Langmuir; Professor Peter Smith and successive epidemiology classes at the London School of Hygiene and Tropical Medicine; Dr. Ekke von Kuenssberg; the late Sir Patrick Laidlaw; Professor John Oxford; and Dr. Andrew Cliff and Professor Peter Haggett.

Professor Negroni very kindly translated Professor Magrassi's Italian paper, discussed in Chapter 4.

Miss Margaret Hope-Simpson and Mrs. Eleanor Hope-Simpson have generously supported the work financially.

Dr. Irvine Loudon prompted the writing of the book and has lent a guiding hand.

Miss Janie Curtis, of Plenum Publishers, has been most helpful.

I am grateful to Professor Golubev of the All Union Influenza Centre, St. Petersburg, for his collaboration and encouragement.

Dr. A. Hill, Dr. Andrew Cliff, Dr. Paul Glezen, Professor John Oxford, Professor Chandra Wickramasinghe, Professor Edwin Kilbourne, Dr. Gillian Air, Dr. Christopher Scholtissek, and Dr. A. S. Beare have kindly permitted us to reproduce their published tables, figures, and portions of the text.

Various publications and publishing houses have also kindly permitted reproduction of material from my own work and from that of other authors.

Contents

1

The Scope and Purpose of the Book

ANTIQUITY OF THE DOUBT THAT INFLUENZA IS CONTAGIOUS

No thoughtful student of the history of influenza can fail to be impressed by how early the concept that the epidemics were being brought about by "contagion" arose, although it is not easy to be certain what process the writers were envisaging by that term before the microbial discoveries. The opinion has gained ground to such an extent that it is now almost universally believed that, like measles, influenza is caused by transmission of the specific parasite directly from the sick patient to produce the disease within a few days in the nonimmune companions whom he has infected. Indeed, this process, which we shall call the *current concept of direct spread*, is now so universally accepted as to seem self-evident, needing no defense.

The task undertaken in this book, namely to advance an alternative theory, which for simplicity we call the *new concept*, is therefore one of some difficulty because a contrary opinion must endeavor to win acceptance by identifying those aspects of the relationship between mankind and the influenza viruses that the current concept is powerless to explain, and then demonstrating that they are explicable by the new concept. Nevertheless, few concepts are ever entirely novel, and students of the history of influenza will also have become aware of a powerful minority view that has been unable to accept the validity of direct contagion. Whether such early opponents were agnostic, confessing their inability to propose an alternative hypothesis, or whether they blamed the malign constitution of a particular season or the baneful effects of miasmata or of earth poisons, they had in common their inability to explain the epidemic process of influenza by direct spread from the sick persons. These minority voices kept recurring throughout the last four centuries, and there is a danger that attempts to paraphrase their writings may unwittingly alter their meaning, so where possible the actual writings have been quoted. To balance the picture, quotations have also been introduced from writers who favor the concept of contagion.

Opinions concerning the nature of infectious diseases tend to be dismissed if they were written before Pasteur, Koch, and others had established that many diseases are caused by direct transmission of specific microorganisms. One must of course allow for the current state of knowledge in interpreting such early observations but they should not be discounted. Many writings on influenza from the seventeenth to the nineteenth century are worthy of study. The disease is often clearly recognizable, the observations acute, and the inferences reasonable. Chapters 2, 3, and 4 chronicle the debates that divided the doctors who believed that influenza was contagious from those who considered that the evidence precluded such an interpretation of the epidemic process.

THEOPHILUS THOMPSON'S ANTHOLOGY

Human epidemic influenza is often so dramatic and causes so much social disruption over vast areas in so short a time that it has left its records in the writings of physicians, clergy, and literate laymen of past centuries. They show that influenza has been with us throughout recorded history as a ubiquitous pestilence that seems to have altered little through the last five centuries.

I am greatly indebted to Drs. J. Allen McCutchan, Arthur Friedlander, Michael Oxman, and Abraham Braude of San Diego for securing me a copy of Theophilus Thompson's *Annals of Influenza in Great Britain from 1510 to 1837.*[1] In 1852 Thompson performed the invaluable service of compiling this anthology on influenza. He explains why he did so in his short introduction. Influenza is, he says;

> . . . of all epidemics the most extensively diffuse, and apparently the least liable to essential modification, either by appreciable climatic changes, or by hygienic conditions under the control of man. It is not like Smallpox, communicable by inoculation; and, however its fatality may be influenced by defective drainage, it is not like Typhus, traceable to the neglect as its cause. Unlike Cholera, it outstrips in its course the speed of human intercourse. It does not, like Plague, desert for ages a country which it has once afflicted, nor is it accustomed, like the Sweating-Sickness, in any marked manner to limit its attack to particular nations, or races of mankind. There is a grandeur in its constancy and immutability superior to the influence of national habits. The changes in our national system of diet during the period which this Volume embraces, have been calculated to effect remarkable modifications in the condition of the people in reference to disease, yet, as respects Influenza, they are not proved to have exerted any manifest influence. The disease, moreover, exhibits in the well-ordered mansions of modern days, phenomena similar to those which it presented in the time when rushes strewed the ground in the presence chamber of our monarchs, and decaying animal and vegetable matter obstructed the porticoes of palaces. (p. xi)

He also wisely remarks in his introduction that:

> No single generation of medical practitioners can be expected to possess a sufficient range of observation, or to accumulate adequate materials of information on the subject, to enable them to detect the clue by which to thread the intricacies of this inquiry. The past must be scrutinised, and its reflected light brought to our aid; old and new facts when collated, by the harmony which they exhibit, become mutually illustrative, and acquire a value previously unknown. It is true, that medical records abound in fallacious and imperfect observations, transmitted from one generation to another, and that popular prejudices have exercised an influence in disseminating error, which the obstinacy engendered by the evidence of imperfectly observed facts has tended to confirm and perpetuate; but it is possible to manifest too indiscriminate a contempt for statements which partake of popular superstition. Popular opinion is not necessarily incorrect, because inconsistent with the views of contemporary philosophers.

The good fellow has endeavored to present the picture of influenza exactly as it is delineated by the original observers in order to avoid coloring it with any bias present in his own mind "since narratives free from all preconceived impressions contribute far more effectually . . . to the formation of clear and independent opinion." I have attempted to follow this example when introducing the opinions and observations of others in the discussions in this volume.

DUBIETY DESPITE ADVANCE IN KNOWLEDGE

Although the book was undertaken in order to advance the claims of one particular concept of the epidemic process in influenza, it has, almost of its own volition, become a miniature history of the rival hypotheses that have been developed. One might suppose that the questions of "contagion" would have been finally settled in 1933 when it was discovered that human influenza is caused by a specific ultramicroscopic parasite. In one sense that was true. There could no longer be any doubt about the agent that was causing the disease. Miasmata and earth poisons could be dismissed. Yet within a few years we find the very persons who had been involved in the discovery of influenza A and B viruses—Richard Shope, Patrick Laidlaw, Christopher Andrewes, Macfarlane Burnet, Thomas Francis, Jr., working in four widely separated parts of the world—independently expressing their perplexity in their attempts to understand how the organisms they had discovered were behaving.

As information about influenza viruses has accelerated, so has perplexity increased. The rapidly expanding knowledge about the parasite has brought with it fresh epidemiological and other problems demanding explanation. The nonspecialist reader is required to hold a mental picture of the structure of the influenza virus and to understand how it behaves and multiplies if he is to appreciate the problems facing the epidemiologists and if he is to enjoy and evaluate the hypotheses and debates that have arisen from their endeavors to explain the findings. A simple account has therefore been provided of the structure of influenza A and B viruses and of their interactions with their human host, necessarily including discussion of their antigenic variations, especially of those of type A influenza virus (See Chapter 5 and 6).

The 1950s saw some remarkable findings. One of the more fortunate accidents of the history of influenza has been the literary skill of the persons who have studied the disease and the felicity and charm of their writings. The reader cannot fail to be struck by the vigor and directness of Thomas Willis and other early writers, and many later writings on influenza are a pleasure to read, from Andrewes, Burnet, Shope, Davenport, and Francis to Stuart-Harris, Kilbourne, and Scholtissek. Many of the writings are so trenchant and enjoyable that there was a temptation to transform the book into an anthology. It has seemed right to present some of the exciting serological discoveries in the words of those who made them. The findings are not yet entirely understood, and the reader should know precisely what those most involved in the discoveries thought about them.

PERSONAL INVOLVEMENT IN THE PROBLEMS OF EPIDEMIC INFLUENZA

My own involvement with the problems posed by epidemic influenza began more than half a century ago during the 1932–33 epidemic during which the first human influenza virus was discovered. The behavior of that epidemic in a rural population in the county of Dorset seemed to be inexplicable in terms of a virus that was spreading directly from the sick in the manner of measles virus. The attempt to find a rational explanation of its behavior has preoccupied me ever since, as it has so many others. The great epidemic of 1951 provided an opportunity for further study of the disease not only in the general practice population but also at the World Influenza Center at the Palais des Nations in Geneva. The Medical Research Council and its Public Health Laboratory Service Board (PHLS) had established a small epidemiological research unit in our Cirencester (Gloucestershire) general practice in 1947 with the object of studying the natural history of the common diseases. Thanks to this generous help it was possible to investigate all the influenza epidemics that attacked the Cirencester community from 1947 to 1976. During the early years it had been necessary to send laboratory specimens to London for virological and serological examination, but in 1961 the PHLS added a virus laboratory and staff directed by Dr. Peter G. Higgins within the general practice premises. A 24-hour daily watch was kept continuously for 15 years on the diseases of the small community of some 4000 persons. After 11 years, Dr. Higgins was succeeded by Dr. George D.H. Urquhart and later still by Dr. Brian Roome.

Some of the findings of the Cirencester surveillance relating to the “Asian” influenza era (1957–1968) and the “Hong Kong” influenza era that succeeded it are included in Chapter 7. The epidemics were studied in increasing detail and the problems that they posed ultimately provided the insights that led to the new concept.

DEFICIENCY OF THE CURRENT CONCEPT: THE INFLUENCE OF SEASON

The more intensively influenza was studied—whether in the field or in the laboratory, within the household or on a global scale—the more aspects came to light that demanded explanation. In 1977, Fred Davenport[2] reminded his colleagues that: "Epidemiological hypotheses must provide satisfactory explanations for all the known findings—not just for a convenient subset of them." His uncomfortable dictum is relevant to our discussion. He was reminding us that a concept of influenza epidemic process resembles a milk bucket. Overlook even a single hole in the bottom of the bucket and all the milk is lost. A solitary aspect of the behavior of human epidemic influenza or of its causal viruses that cannot be explained by the concept is just such a hole in the bucket. It invalidates the concept unless some adjustment can mend it. So many holes exist in the current concept of direct measles like spread of influenza that the hypothesis more resembles a colander than a bucket. It cannot, for example, offer any explanation of antigenic variation, the explosive nature of many influenza epidemics simultaneously attacking vast populations over wide areas, the cessation of epidemics where they are admirably placed to continue by direct spread, and numerous other difficulties that will appear later.

The most striking defect of the current concept of direct spread is its inability to explain a common feature of human epidemic influenza, namely the fact that it is a seasonal disease. This is so well known as to be taken for granted. Perhaps, for that very reason, it has tended to be forgotten that, as an aspect of influenza behavior, it needs to be explained by any valid concept of the epidemic process. No reference to season can be found in the index to the 1975 edition of *The Influenza Viruses and Influenza* edited by Edwin J. Kilbourne.[3] Half a century earlier, Wade Hampton Frost,[4] whom many consider to have been the father of modern epidemiology, had devoted much attention to the seasonal nature of influenza and other diseases, but had failed to find an explanation of the seasonal character of influenza. He had, however, realized the importance of the phenomenon in relation to the concept of the epidemic process:

> The seasonal fluctuations in rates of prevalence which are characteristic in many diseases can usually be explained, if at all, only in the light of fairly definite knowledge of other associated epidemiological features; hence, considered by themselves, these fluctuations must be interpreted most cautiously. The seasonal distribution may, however, support or negate a particular hypothesis as to the means of spread of a disease of uncertain epidemiology.

We, too, had been omitting the seasonal influence from our attempts to explain the epidemic process in influenza, and its subsequent inclusion was largely responsible for generating the new concept. The influence of season on influenza is so important that it has merited a separate chapter. Chapter 8 gives a brief résumé of the solar–terrestrial relationship that, however remotely mediated, causes seasons and all seasonal phenomena, including seasonal diseases. This relationship illustrates an immutable natural law, and Chapter 8 shows how it was tentatively integrated into the new concept and how the integration empowered the concept to explain many and possibly all the difficulties inexplicable by the current concept. Examples are given of the seasonal behavior of influenza both locally in small communities and globally. The new concept suggests a different epidemic mechanism whereby the apparent movement of epidemic influenza does not reflect the movement of the virus from person to person. The movement seems to be reflecting the regular seasonal journey of an unidentified stimulus dependent on the seasonal variations in solar radiation that determine all seasonal phenomena. A "prediction" can therefore be made retrospectively, namely that the speed of travel of influenza cannot have altered throughout the centuries despite the alteration in speed and complexity of human communications. Evidence is given supporting this "prediction" (see Chapter 17).

If the new concept is correct, it suggests an approach to other common respiratory diseases whose seasonal epidemiology has hitherto proved intractable.

THE NEW CONCEPT OF VIRUS PERSISTENCE, LATENCY, AND SEASONAL REACTIVATION

The new concept was first published in 1979 in the *Journal of Hygiene* (now *Epidemiology and Infection*) and further evidence in support appeared in successive papers in that journal during the next seven years. In brief, the concept proposes that influenza virus is seldom transmitted from the human host during the influenza illness because it too rapidly adopts a persistent noninfectious mode. The ex-patient carries this persistent influenza infection for a year or two, but it is reactivated annually by a seasonally mediated stimulus. Epidemics consist of the nonimmune companions of such carriers who have been infected during the brief periods of high infectiousness at the seasonal reactivation.

Carriers themselves seldom suffer any ill effect from the presence of their persistent colonies and do not often have a further attack of influenza during the reactivation. After a year or two the persistent infection terminates in a more or less permanent parasitism of the carrier's respiratory epithelium by the genomes of the virus. It is proposed that seasonally mediated reactivation of these genomes is concerned with the phenomena of antigenic shift of influenza A virus and recycling of its major serotypes. This lifetime harborage of the influenza A virus genome and subsequent reactivation may apply only to the first ever influenza A infection of a person's life.

Chapter 16 sets out the new concept in 11 detailed propositions, each of which is discussed. The application of the hypotheses to explain the many difficult features of the behavior of influenza and its causal viruses is a constituent of many chapters. Chapter 9, for example, discusses the nature of antigenic drift and the difficulties that it presents for the current hypothesis of direct spread. The new concept provides a relatively simple explanation supported by laboratory experience of the production of antigenic drift.

Antigenic shift of influenza A virus offers a number of interesting and difficult epidemiological problems. These are discussed in Chapters 10 through 12 in relation to various hypotheses including those of the new concept.

The new concept proposes that the influenza virus adopts two modes of parasitism in the human host, persistence and latency, which have not yet been demonstrated in human influenzal infection. Chapter 13 therefore reviews the potential of influenza A virus to adopt a variety of natural relationships with nonhuman hosts and Chapter 14 extends the investigation to the behavior of the virus in the laboratory. Such influenzal parasitism of nonhuman hosts is of great interest and suggests possible explanations of the complex behavior of the parasite in mankind. We therefore discuss the lively debate as to the place of mammalian and avian influenza in human influenza epidemiology.

OTHER ALTERNATIVE HYPOTHESES

These and the subsequent chapters attempt to present the reader with a comprehensible picture of the epidemic process by which the influenza virus has survived successfully as a parasite of the human population of the world for the last 100 years, interpreting the phenomenon in the light of the new concept and discussing alternative hypotheses that have been proposed.

At the beginning of our narrative in Chapter 2, Thomas Willis is quoted as describing the 1658 influenza epidemic as follows: "About the end of April, suddenly a Distemper arose, as if sent by some blast from the stars, . . . " so rapidly did it seize hold of large numbers of people. We are not to take his famous simile as evidence that Willis believed that the epidemic had actually been caused by such a terrestrial invasion from space, but Chapter 12 contains a description of such a postulate made more than 300 years later by two nonmedical scientists attempting to explain epidemics in the twentieth century.

The hypotheses advanced in the new concept are likely to be superseded in part or altogether as more information is gathered. This is the destiny of all hypotheses. It seems certain that the current concept of direct spread is impeding our understanding of influenza. The later chapters of this book give a quite different picture of what has been happening, and the new concept will serve its purpose if it acts as a platform from which greater understanding can be reached.

THE IMPORTANCE OF THE HISTORY OF THE BEHAVIOR OF HUMAN INFLUENZA VIRUSES

When the first major change in the serotype of influenza A virus was noticed in 1946, it was naturally assumed that the human population had been invaded by a novel virus strain. The conclusion was reinforced when in 1957 another major variant of influenza A virus replaced the 1946–1955 novelty, and microbiologists began speculating as to whence these new viruses were being derived. There was much surprise many years afterward when serological studies produced evidence that all three subtypes of human influenza A virus had probably had previous eras of prevalence long before the discovery of the virus in 1933.

The history of these findings is related in Chapters 3, 6, 10 and 14, and they provide yet another obstacle to accepting the current concept of direct spread of influenza and provide support for the new concept.

The book has of necessity approached the problems of influenzal epidemiology from many different directions, and it attempts to assemble the evidence for the new concept while not concealing the problems that have yet to be solved.

REFERENCES

1. Thompson T: *Annals of Influenza or Epidemic Catarrhal Fever in Great Britain from 1510 to 1837*. London, The Sydenham Society, 1852, pp ix–x.
2. Davenport FM: Reflections on the epidemiology of Myxovirus infections. *Med Microbiol Immunol* 164:69–76, 1977.
3. Kilbourne ED: *The Influenza Viruses and Influenza*. New York, Academic Press, 1975.
4. Frost WH: Epidemiology, in Maxcy KF (ed): *Papers of Wade Hampton Frost*, New York: The Commonwealth Fund, 1941, pp 520–521. (First published in the Nelson Loose-Leaf System, *Public Health-Preventive Medicine*, 2:163–190, 1927. New York, T. Nelson & Sons.)

2

The Debate about the Contagiousness of Influenza

THE UBIQUITY OF INFLUENZA

Communities of human beings have established themselves throughout the habitable regions of the globe, and, apart from some very isolated pockets of humanity, they have experienced outbreaks of influenza wherever they were living between the North and South Poles. Influenza has been characterized by this ubiquity for at least 400 years and possibly for far longer. One cannot be certain of the nature of the many earlier pestilences recorded in contemporary annals but it is probable that some were influenzal. Medical historians have begun to feel confident that vivid accounts of "epidemic catarrhal fever" and other more fanciful diagnoses in the seventeenth and early eighteenth century describe epidemic influenza. The name "influenza" was first used in England to describe the influenza epidemic of 1743.

AN INFLUENZA EPIDEMIC IN THE SEVENTEENTH CENTURY

Throughout the centuries observers have commented on the abruptness of the onset of some influenza epidemics and the simultaneity with which the epidemics have attacked populations covering large areas. For example, Thomas Willis,[1] wrote the following dramatic description two months after the beginning of the explosive influenza epidemic of 1658:

> about the end of April, suddenly a Distemper arose, as if sent by some blast from the stars, which laid hold on very many together: that is some towns, in the space of a week, above a thousand people fell sick together. The particular symptoms of this disease, and which first invaded the sick, was a troublesome cough, with great spitting, also a Catarrh falling down on the palat, throat and nostrils; and also it was accompanied with a feverish distemper, joined with heat and thirst, want of appetite, a spontaneous weariness, and a grievous pain in the back and

> limbs. . . . loathing of food, . . .But in some a very hot distemper plainly appeared, that being thrown into bed they were troubled with burning thirst, waking, hoarseness, and coughing almost continual; such as were induced with an infirm body, or men of a more declining age, . . . not a few died of it. . . . Concerning this disease, we are to inquire, what procatartic [sic] cause it had, that it should arise in the middle of Spring suddenly, and that the third part of mankind almost should be distempered with the same, in the space of a month.

No physician who has had to deal with a severe influenza epidemic can doubt that Willis is describing one. He outlines problems that still confront us: the explosive onset, simultaneous among many people, and the brief duration of the epidemic.

IDEAS ABOUT INFLUENZA IN THE EIGHTEENTH CENTURY

The eighteenth century witnessed a surge of interest in epidemic influenza. The teachings of Hippocrates and Galen and the writings of Thomas Sydenham were exerting a profound influence on the concepts of physicians, so that they were looking into the peculiar "constitution of the season" for the explanation of the origin of an epidemic or seeking it in the meteorological conditions preceding the outbreak. The influence of such teaching has not altogether disappeared, but even then doubts were stirring, and many physicians, reflecting on their experience of the disease, expressed dissatisfaction with the theories then current. Here is Robert Whytt[2] of Edinburgh discussing the influenza epidemic that had attacked the United Kingdom in the autumn of 1758:

> I thought it proper to lay before you this account of the weather in order to judge how far any sensible changes of the air might influence the health of the people here. But for my part, considering how remarkably mild and dry our season was, I can hardly ascribe the rise of our epidemic to any of the known qualities of the air.

Dr. John Millar[3] of Kelso, Scotland, on the other hand, favored a meteorological cause because direct infectiousness seemed to him to be an inadequate explanation of the 1758 epidemic:

> Slight colds generally come on after the autumnal equinox but there are few instances of any that have prevailed so universally as the epidemic cold which has raged here for these two months past. It did not seem to be produced by any other contagion than that of the air, because all the same family that were seized with it generally fell down at once, and those who escaped at its first entrance, were not afterwards affected: nor did it spread, as might have been expected, were it infectious.

Students of the epidemiology of influenza owe a debt of gratitude to a remarkable eighteenth-century physician, Dr. John Fothergill. He had been one of the early medical students at the University of Edinburgh's medical school which had recently been launched on the initiative of a group of Scottish doctors trained at Leyden University under the stimulating influence of Boerhaave. John Fothergill never lost that curiosity for inquiring into the causes of natural phenomena by examining them directly, not too much prejudiced by the teaching of the ancient masters.

Besides being an assiduous student, Fothergill was friendly and sociable. He was instrumental in starting the first association of medical students in Edinburgh. The friendships he made among fellow students, teachers, and successors were lifelong. He set up in medical practice in London and became the foremost physician in the country despite his modest disposition. He kept up an extensive correspondence with his medical friends and his aid and advice were sought by subsequent generations of fledgling doctors.

Consequently, when he was intrigued by the epidemic mechanism underlying the sudden widespread appearance of mild influenza in the last months of 1775, he was able to call into being a network of medical observers in many different places in England, Wales, Scotland, and Ireland. He seems to have been the first person to have used the technique of a more or less formal questionnaire. He wrote a sketch of the 1775 influenza as he had experienced it among his own patients, and he sent a copy together with a list of questions to about 20 colleagues practicing in different parts of the four countries.

Like his friend Sir George Baker, Fothergill was inclined to mistrust speculation: "though attempts to ascertain the causes of epidemics are, for the most part, more specious than substantial . . . " he nevertheless requested the opinions of his correspondents about the cause of the epidemic, and he provided a detailed description of the "constitution" of that autumn season. He called for an opinion on the contagiousness of the disease though he did not discuss it in relation to humans, but he noted that both horses and dogs were much affected by a similar disease, "those especially that were well-kept."[4]

Another old friend, Sir John Pringle, commended Fothergill for his meticulous analysis of the meteorological situation, but immediately dismissed such an approach as not possessing any value toward understanding the etiology of influenza. He pointed out that the 1775 epidemic had afflicted numerous parts of Europe in which the meteorological conditions had differed markedly from those in the United Kingdom. He concluded that " . . . such epidemics do not depend on any principles we are yet acquainted with, but upon some others, to be inquired upon."[5]

Other respondents to Fothergill's questionnaire differed from one another on the important question about the infectiousness of the disease. Dr. D. Campbell of Lancaster, for example, considered that the progress of the epidemic northward from London and a number of similar observations had fairly proved its contagious nature.[6] Dr. Thomas Glass[7] of Exeter, on the other hand, thought it could not be so:

> Nor does this distemper arise, which is, I think, at present, the more general opinion, from contagion. For in this city, in the year 1729, it was conjectured, that two thousand persons at least were seized with it in one night. But what is more extraordinary, before the Autumn, in the year 1557, it attacked all parts of Spain at once, so that the greatest part of the people in that Kingdom were seized with it almost on the same day. This very singular circumstance is related by Mercatus, who says it happened in his own time.

Singular indeed! Glass's subsequent suggestion is perhaps more worthy of consideration, namely that the pestilence described by the ancient Greek poet Homer may have been epidemic influenza, which:

> . . . within the space of nine days, spread itself over all the Grecian quarters, and a little while after disappeared, was an epidemic of the same kind; because neither the true plague, nor any other epidemical disease, with whose history I am acquainted, has been known to make so rapid a progress, or to end so soon, as that pestilence did.

One should note the astute observations of the epidemiological significance of the explosive onset, brief presence, and abrupt termination of some severe influenza epidemics. Later in this book we have to consider the mechanisms that underlie such characteristics.

The remarks of Dr. John Haygarth[8] of Chester in the county of Cheshire on the border of North Wales are of particular value because of the care he took in attempting to obtain accurate answers to Fothergill's questions:

> The epidemical catarrh of 1775 seized, in general, the inhabitants of Chester about the middle of November. From the 15th till the 25th of that month the distemper spread most universally; yet very few were attacked so late as December. Indeed I saw one case on the 2nd of November, of a lady who had suffered manifest symptoms of this epidemic six days before [i.e., 26 October]; but I heard of no other instance of its appearing here so early, and the disorder did not become general till near a fortnight later. This epidemic pervaded all North Wales within three or five days after its general seizure of the inhabitants in Chester; that is, on the 18th or 20th of November, as I have authentic information from every town and every considerable village, and their neighborhood. I was curious to know how those were affected who were most secluded from the intercourse of society; an intelligent practitioner informs me that in Llyn, the most western and remote corner of Carnarvonshire, this epidemic began about the 20th November, was general through every part of this peninsula, and affected all classes of people; that one in a family now and then escaped it, but that he knew no family, however small, among whom it did not make its appearance. My medical correspondents mention that some cases occurred in one part of Wales so early as October the 27th, and in another the beginning of November. In the western part of Cheshire [bordering Wales], and that part of Shropshire which borders on Cheshire, I observed that this disease began soon after the middle of November. However, I am certain that in some Cheshire villages the epidemic had not appeared till more than ten days later, though it afterwards visited these places. These facts, compared with the general seizure, make the theory of this epidemic very difficult. On the whole, I believe people in the country were attacked rather later than in the towns they surrounded, less severely and less generally; however, not only the inhabitants of villages, but of solitary houses, were seized with this disease. I could not discover that high or low, dry or moist situations, the neighbourhood of mountains, or of the sea, or any other particular exposure, rendered the epidemic either later or milder; though I made very circumstantial inquires to ascertain these facts.

Haygarth is evidently puzzled how to explain the features of the 1775 epidemic that he has studied so carefully. This was not his only contribution to the subject as we shall see.

The example of Dr. Fothergill's surveillance was soon followed by others. Dr. Lettsom, a protégé of Fothergill who had been born on a Caribbean island, had founded a society in London for promoting medical knowledge that later became the London Medical Society, a distinguished body that still flourishes. Fothergill was a foundation member. The next considerable influenza epidemic is recorded as occurring in March and April of 1782. The London Medical Society requested Dr. Edward Gray, who had been a fellow student with Fothergill at Edinburgh, to conduct a surveillance similar to that inquiring into the 1775 epidemic.

Dr. Anderson, writing from Alnwick, Northumberland, replied that: "... with regard to the number affected, it [the 1782 outbreak] was the most universal disease ever remembered."[9] This opinion was echoed by Dr. Murray[10] from Norwich, Norfolk, and by Dr. Kirkland[11] of Ashby-de-la-Zouch, Leicestershire. Dr. Gilchrist[12] of Dumfries in Scotland expressed the contrary view that it had not attacked so many people as the 1775 epidemic.

Dr. Ruston[13] like Dr. Glass, an Exeter physician, made the percipient observation: "... it was so universal that it may rather be said to have ceased for want of subjects, than to have lost the power of exerting its deleterious effects."

The remarks of Dr. Binns[14] of Liverpool apply equally to our own experience 150 years later in the 1932–33 influenza epidemic: "Whole families were affected by it at the same time, so that no one remained well to nurse the sick; and it was extremely difficult to get any assistance, as none remained free from the disease."

Small children, the elderly, and persons who had been attacked in 1775 are noted as tending to have escaped in the 1782 epidemic.

Dr. Gray[15] discussed the opinions of his respondents as to the manner in which the 1782 epidemic had arisen and been propagated.

> Some physicians thought it arose solely from the state of the weather ... by changes in the sensible qualities of the atmosphere, such as increase of cold, or moisture ... unconnected with any disorder that had prevailed, or did at that time prevail, in any other part. Others, admitted its cause to be a particular and specific contagion, totally different from, and independent of, the sensible qualities of the atmosphere, yet thought that cause was conveyed by, and resided in the air. But the greatest number concurred in the opinion, that the influenza was contagious, in the common acceptation of that word: that is to say, that it was conveyed and propagated by that contact, or at least by the sufficiently near approach, of an infected person.

Despite dissenting voices, Dr. Gray cast his own vote in favor of the last hypothesis of direct contagion, which is still the current concept.

A committee of Fellows of the Royal College of Physicians of London also collected observations about the 1782 influenza epidemic and published its findings during the following year. A conflict of evidence between the two investigations well illustrates the difficulty that still attends such field studies.

Concerning the epidemic as it affected the city of Newcastle-upon-Tyne, Dr. Gray had quoted a letter from Dr. Clark of that city that the disease had appeared at Shields, the port of Newcastle, on or about 20 May after some ships had arrived in the port from London where influenza was already epidemic, the crews having suffered influenza on the voyage. Dr. Clark stated that the first Newcastle family to be attacked was seized on the 28 May, and that they kept a public shop that probably served such sailors. Other evidence is adduced to support this plausible picture of the disease being introduced into Newcastle by infectious persons bringing the parasite by sea from London.[16]

The report to the Royal College of Physicians tells a different story: "The earliest intelligence given of the disease is, that it appeared in Newcastle-upon-Tyne in the latter end of April 1782, and raged there during the whole month of May, and part of the month of June."[17] On this evidence London might be supposed to have been infected from Newcastle.

The college report also contains the famous account of the impact of the 1782 influenza on Admiral Kempenfelt's squadron of warships. The squadron had sailed from Spithead on 2 May and had been cruising between Brest in France and the Lizard Peninsula in Cornwall without having any communication with the shore. The crew of *HMS Goliath* was attacked by influenza on 29 May after about four weeks at sea and the other ships thereafter, and so many sailors were laid low that the whole squadron was compelled to return to port during the second week of June.[18]

Such observations have great importance. It is difficult to see how *HMS Goliath* can have been infected by direct spread of the agent from sick persons.

When we last encountered Dr. John Haygarth, he was perplexed in his attempt to understand the spread of the 1775 influenza epidemic despite wide and careful investigation of it. His subsequent experience during the 1782 epidemic seems to have convinced him of its contagious nature. He communicated his findings to the college but they were not published at that time. More than 20 years later he published them under the title "Of the Manner in Which the Influenza of 1775 and 1782 Spread by Contagion in Chester and its Neighborhood." He then writes:

> Why the publication of this disquisition should have been delayed for twenty years, and yet why it is now laid before the public, may require some explanation.
>
> The contagious nature of the influenza had, I thought, been sufficiently proved by many physicians, and among others by Dr. Falconer, in his account of the epidemic of 1782 ["Treatise on Influenza" by Broughton and Falconer].
>
> But a contrary and, as I think, a very pernicious opinion has lately been supported by physicians of great respectability, and authors of the highest reputation, not, indeed, in this, but in other enlightened nations, have ascribed not only this but many other epidemics, even the plague itself, to a morbid constitution of the atmosphere, independent of contagion. To determine whether this doctrine be true or false, is of the highest importance to mankind. Knowledge, in this instance, is power. So far as it can be proved, that a disease is produced by

contagion, human wisdom can prevent the mischief. But the morbid constitution of the atmosphere cannot possibly be corrected or controlled by man.[19]

He then discussed the epidemics of 1775 and 1782 and compares them with what happened in a third epidemic in 1803, the year in which he was writing. He tackles the questions that had been circulated by Fothergill: questions 1, 2, and 3: How far does the propagation of influenza depend on the climate, weather, or season? He dismisses all three. Questions 4 and 5: Is it conveyed by the wind, or does it spread like sound from a center uniformly and gradually to surrounding places? No! Question 6: Does the first patient contaminate the atmosphere so as to render the place generally pestilential? No, the picture does not fit the hypothesis. Question 7: Does it spread by contagion? Haygarth writes:

Many facts above related manifestly proved the truth of this conclusion. At Chester and most of the towns which surround this city, I had the good fortune to discover the individual person who brought it into each place, previous to the general seizure of the inhabitants.

He supports the statement with a table (Table 2.1) obtained from his own questionnaire among his neighboring colleagues. Anyone who has attempted to carry out this sort of field epidemiology will be aware of the difficulties attaching to it and the caution that must be exercised in assessing the results. Suppose, for example, that the first patient in Chester had not been suffering from influenza. Suppose that the first case had not been attended by doctors.

Nevertheless Dr. Haygarth was a careful and thorough student and he gives details of the first case in Chester:

In 1782, a gentleman ill of the influenza came from London to Chester on the 24th of May. A lady, into whose family he came, and to whom he is since married, was seized with the distemper on the 26th of the same month.

TABLE 2.1. Date on Which the 1782 Influenza Epidemic Is Recorded as Beginning in Chester and Neighboring Towns[a]

Town	Distance (miles)	First patient	Days after Chester
Chester	—	May 26	—
Tarporley	SE 10	June 6	11
Holywell	NW 18	June 6	11
Malpas	SE 15	June 7	12
Frodsham	NE 10	June 7	12
Middlewich	E 20	June 9	14
Wrexham	SW 12	June 10	15
Mold	W 12	June 13	18
Ruthin	W 20	June 14	19
Oswestry	SW 28	June 14	19

[a]Adapted from Haygarth.[19]

The question that provoked the replies that have been copied in Table 2.1 would nowadays be considered to have been "loaded":

> As the first patient I had seen in the influenza of 1775 was the landlady of a principal inn, and as I had observed so distinctly that the epidemic of 1782 was brought into Chester by a patient coming from London, I stated this question to my correspondents: "Could you discover whether the distemper was introduced into your town from any place where it had previously attacked the inhabitants?"

He received confirmation of his suggestion from physicians in all the towns in the table except Holywell and Ruthin. His conclusion, though reasonable, does not, in fact, follow from his findings. He says:

> The intercourse is greater from the metropolis to Chester than to the other towns in its neighbourhood. Again, more people go from Chester to the adjacent market towns than to the villages and scattered houses which surround them. [We have no means of checking this statement]. The influenza spread exactly in this order of time, from the metropolis to Chester, to the neighbouring towns and lastly to the villages.

According to his own earlier statements, the process was different. The influenza at Frodsham had come from Manchester; at Malpas we are only informed that the first case was in the landlady of an inn there; the Middlewich outbreak is said to have been imported from Liverpool; the origin of that at Oswestry is unstated; Tarporley derived it from Warrington via a postillion. Only Mold and Wrexham are stated to have obtained their first influenza patients from Chester.[19]

THE NINETEENTH CENTURY BEFORE PASTEUR

Other observations on the 1803 influenza epidemic show that the conflict of opinion had not yet been resolved, though a majority of medical men probably agreed with Mr. Constance[20] of Kidderminster:

> Whether the influenza has been a contagious disease or not I really cannot satisfactorily determine. I am of Sir Roger de Coverley's opinion, when he so wisely decided, on another occasion, "that much might be said on both sides": however, upon the whole, I think that preponderance of probability is in the scale of contagion.

No statistician could ask for more than a preponderance of probability. However, a distant predecessor of my own practice in Cirencester, Gloucestershire, Mr. Lawrence, was not convinced. If infectiousness is occurring, he thought it must be weak.[21] The late Dr. Marguerite Pereira made a similar observation to me 173 years later when she was studying influenza among some London families, but most of us would find it incredible that influenza is only weakly infectious. The new concept suggests that it is highly infectious but only from symptomless carriers.

Thompson (1852; see note in References) has collected numerous statements concerning the 1803 epidemic mostly culled from volume 10 of the *Medical and Physical Journal* and from volume 6 of the *Memoirs of the Medical Society of London*. He has a section of the anthology headed "Contagion" to which he prefaces (p. 222):

> There is no department of the subject regarding which there is so great a diversity of opinion among observers, as on the much vexed question of contagion: it seems, therefore, desirable to present the most definite statements made on each side of the question.[22]

The majority of his quotations favor the concept of contagion, but he comments that many observers had expressed themselves as undecided.

The Council of the Provincial Medical Association, an association of doctors that subsequently became the British Medical Association, circulated a questionnaire to its members after the influenza epidemic of 1836–37. Among the 18 specific questions, the following particularly concern us:

> 12. Are you in possession of any proof of its having been communicated from one person to another?
> 14. Were there any circumstances that appeared to exempt individuals from an attack of the disease? and, in particular, did they, having been attacked during the last similar epidemic of the year 1834, appear to afford any protection?
> 18. Did any peculiar atmospheric phenomena precede or accompany this epidemic?

The answers to the first three questions are also of permanent interest:

> 1. When did the Influenza appear in your neighborhood and how long did it prevail there?
> 2. Did it attack a great many individuals at the same time?
> 3. Did it appear partial to any age, sex, or temperament or did it appear to attack all indiscriminately?

The replies, though not so numerous as had been hoped, came from nearly all parts of the kingdom, and were discussed at the annual meeting of the association at Cheltenham in July 1837. A committee was appointed, the report being written by Dr. Robert J.N. Streeten.[23] It was published in 1838 in the association's transactions, volume 6.

The answers to the above questions were broadly as follows:

1. RE: Question 1: Dates of onset and duration are widely discrepant, though all reported the greatest prevalence from mid January to the end of the first week in February 1837.
2. RE: Question 2: The replies are

> uniformly in the affirmative, and by far the greater portion of them speak decidedly as to the simultaneous outbreak of the disorder throughout the localities to which they severally refer. Dr. Davis, of Presteign, observes, that within his district comprising a circle, the diameter of which is about fourteen miles, it was impossible to make any progression—cases in every part of it occurring simultaneously. Mr. May, of Reading, and several other gentlemen make the same remark.

Dr. Shapter of Exeter, Mr. Bree of Stowmarket, Mr. Maul, and others noted sporadic cases preceding the attack. A very great prevalence of the epidemic had evidently occurred in all parts of the kingdom:

3. RE: Question 3: More than half of the respondents stated that the attacks occurred indiscriminately at all ages and in all sorts of persons. A considerable number of observers, however, agreed that children under six years of age were selectively spared. A careful report from Chichester stated:

   it seems almost equally to have attacked young and old. Of cases recorded, the greater number appear to be at the periods under ten, and from thirty to forty, but the difference in the intermediate decades was trifling, and the uniformly decreasing numbers beyond forty would probably about tally with the small population in those ages.

   The sexes seem to have been equally susceptible.

4. RE: Question 12: The answers to the question about communicability are surprising, being almost uniformly in the negative sense:

   . . . the opinion of nearly all those who had the most extensive opportunities of investigating the disease, and best means at arriving at a definite conclusion, being, that there is no proof of the existence of any contagious principle by which it was propagated from one individual to another.

   The Chichester report, in which all the doctors of that place collaborated, stated:

   We have no proof of the disease having been communicated from one person to another, though the patients often suspected it themselves. Our observations, however, incline us to the opposite belief. It was no uncommon circumstance for the persons who had nursed a number of influenza patients to escape it themselves entirely.

   Dr. Streeten gives five instances of returns that conflict in some degree with this negative opinion but concludes that "nothing approaching tangible evidence [of communicability] is afforded by any of these statements.

5. RE: Question 14: The greater number of respondents "agree in stating that their having undergone an attack in the previous epidemic [of 1834] afforded no protection, and that there were no circumstances which appeared to exempt from an attack of the disease." Some practitioners thought that sufferers from the 1834 epidemic were more liable to attack in 1836–37. A few doctors, however, gave the opposite opinion that the earlier epidemic had afforded protection to its sufferers against the 1836–37 epidemic.

   If the conclusion is correct that the 1834 epidemic afforded no protection, there must be a strong suspicion that the 1836–37 epidemic marked the occasion of an antigenic shift or that one epidemic was of type

A and the other of type B influenza. Theophilius Thompson (1852, p. 340) noted the presence of the epidemic in Australia, South Africa, the Baltic countries, and the north of Scotland, bespeaking a pandemic distribution. The impact that it had on the general mortality confirms it among the severe visitations of influenza.

Thompson (1852, p. 359 *et seq.*) includes, toward the end of his anthology, an account of the impact of the influenza of 1836–37 on the British naval ships of war, both in their home and foreign stations and at sea. These detailed reports would seem to suggest to the reader that a communicable agent was responsible, although Dr. Thompson regards them as not incompatible with some meteorological influence. He is candid about the difficulty of deciding the issue and urges the reader to keep his mind open to many other possibilities, some of those which he suggests seeming bizarre to modern opinion.

OPINION IN THE NINETEENTH CENTURY AFTER PASTEUR

All the discussions and observations considered earlier in this chapter took place before Pasteur, Koch, Lister, and others had established that specific microorganisms cause specific maladies in plants, humans, and other animals. Their discoveries caused a reorientation in medical opinion in favor of the direct infectiousness of many diseases, including influenza. The hunt was on for a particular parasite as the cause of each acute disease, and when Pfeiffer wrongly identified a tiny bacterium as the influenza bacillus, the dispute seemed to have been settled. Pfeiffer's bacillus was named *Haemophilus influenzae*, and epidemics of influenza were thought to be caused by its direct transmission from the influenzal patients. Nothing could be simpler.

It is therefore remarkable that August Hirsch, the outstanding medical historian and epidemiologist of the latter half of the nineteenth century, found himself unable to accept this opinion. In 1883 his great work translated from the German by Charles Creighton appeared under the title *Hirsch's Handbook of Geographical and Historical Pathology*. The first of the three volumes begins with a study of epidemic influenza. Under "Alleged Contagiousness" Hirsch[24] writes:

> In more recent times the great majority of observers have answered it decidedly in the negative, not so much on the many single observations which tell against the communicability of the disease, as on the ground that the spread of influenza can be shown to have taken place quite independently of intercourse. To this argument I may add that it has not spread more quickly in our own times, with their multiplied and perfected ways and means of communication, than in former decades or centuries. . . . Partisans for the spread of influenza by contagion have found support for their views in the breaking out of the disease at various places somewhat removed from the track of commerce, after the arrival of strangers. . . . Without questioning the accuracy of the observation itself we may hesitate to accept the conclusions drawn from it when

we duly keep in mind that the suspected importers of the morbid poison [the causal microorganism] remain, as we are expressly told, unaffected by it, that they continue untouched by the epidemic, and, further, that the disease has not unfrequently appeared in these [Iceland and Faroë] and other islands at the time of the ship's arrival although influenza had not been prevailing as an epidemic anywhere else, and most certainly not in those countries from which the ships had sailed. These considerations, taken along with peculiarities in the incidence and course of influenza epidemics—their occurrence suddenly and without prelude, and their attacking people *en masse*, their equally sudden and complete extinction after a brief existence, generally of two to four weeks, and the frequent restriction of the disease to one place while the whole country round has been completely free from it—all these points are so foreign to the mode of development and the mode of spreading proper to such maladies as originate beyond doubt through communication of a morbid poison, that we shall find it hard to discover any reason for counting influenza among the contagious or communicable diseases.

Although Hirsch is here pointing out the epidemiological difficulties that make it impossible, without major modifications, to accept a hypothesis of direct spread of the disease, it must not be thought that he is ruling out a microbial causation for influenza. Earlier in the book Hirsch[24] had written:

Influenza is a *specific infective disease* like cholera, typhoid, smallpox and others, and it has at all times and in all places borne a stamp of uniformity in its configuration and in its course such as almost no other infective disease has. Its genesis presupposes, therefore, a *uniform and specific* cause, the origin and nature of which are still completely shrouded in obscurity. (p. 34)

He raises no objection to calling this specific cause by the name "an influenzal miasma," or presumably by any other name, so long as we realize that, by substituting a name for an obscure conception, "we do not bring ourselves by that means a single step nearer to a knowledge of the cause of the disease."

Hirsch disposes of the theory that had recently been received, namely that influenza epidemics are caused by a microbe being carried about by the atmosphere, *a miasma vivum*. He had already established that influenza is independent of the state of the weather and also of the race of the persons attacked, but he is careful to note the many instances when natives indigenous to a country had been attacked by an influenza epidemic that spared the resident foreigners.

His translator, Creighton,[25] who himself wrote a classic of epidemiology, *A History of Epidemics in Britain*, was another distinguished medical scholar who rejected the contagiousness of influenza. He was, however, totally opposed to the theory that diseases are caused by parasitic microbes, so that his dismissal of the direct contagiousness of influenza was a manifestation of his much wider dissent from popular medical opinion.

We had to wait another half century before the correct "uniform and specific cause" of epidemic influenza predicated by Hirsch was discovered, and we shall see that the debate about contagiousness that had divided medical opinion for at least three centuries was not thereby settled. The penetrating observations of Hirsch, Streeten, Haygarth and others were hampered because, though a mechanism of direct spread ran into difficulty in explaining the epidemic process, the only available alternatives seemed to be outmoded theories dating from Hippocrates and Galen or wildly improbable ones involving insects, crustacea, or Saharan or volcanic dust.

REFERENCES

Note: Most of the references in this chapter are taken from Theophilus Thomspon's *Annals of Influenza in Great Britain 1510 to 1837*. London: The Sydenham Society, 1852. The page numbers below refer to this book where the original references may be found.

1. Willis T: The Description of a Catarrhal Feaver Epidemical in the Middle of the Spring in the year 1658, pp. 11–12.
2. Whytt R: An Account of an Epidemic Distemper at Edinburgh, and several other parts in the South of Scotland, in Autumn 1758, p 61.
3. Millar J: An account of the Epidemic. Letter to Dr. Pringle dated Kelso, Dec. 8, 1758, p 66.
4. Fothergill J: A sketch of the Epidemic Disease which appeared in London towards the end of the year 1775, pp 86–89.
5. Pringle Sir John: Letter to Fothergill, pp 89–90.
6. Campbell D: Letter to Fothergill, p 113.
7. Glass T: Letter to Fothergill, p 102.
8. Haygarth J: Letter to Fothergill, pp 108–111.
9. Anderson Dr: Letter to Dr Edward Gray, p 118, footnote.
10. Murray Dr: Letter to Dr Edward Gray, p 118, footnote.
11. Kirkland Dr: Letter to Dr Edward Gray, p 118, footnote.
12. Gilchrist Dr: Letter to Dr Edward Gray, p 118, footnote.
13. Ruston Dr: Letter to Dr Edward Gray, p 120–121, footnote.
14. Binns Dr: Letter to Dr Edward Gray, p 121, footnote.
15. Gray: An Account of the Epidemic Catarrh of the Year 1782, pp 117–148.
16. Clark Dr: Letter on the Influenza, from Dr. Clark of Newcastle to Dr. Leslie, p 145 footnote.
17. An account of the Epidemic Disease, called the Influenza, of the year 1782, etc. by a Committee of Fellows of the Royal College of Physicians in London, 1783, pp 155–164.
18. Influenza in Admiral Kempenfeld's Squadron, pp 157–158.
19. Haygarth J: pp 191–198.
20. Constance Mr: p 233.
21. Lawrence Mr: p 233.
22. Thompson Dr T: p 222.
23. Streeten Dr R J N: pp 292–333.
24. Hirsch A: *Handbook of Geographical and Historical Pathology*, vol 1. London, New Sydenham Society, 1883, pp 34, 36–37.
25. Creighton: *A History of Epidemics in Britain* (2 vols). Cambridge University Press, 1891.

3

Epidemic Influenza, 1900–1932

1900–1916

There are several periods in the history of influenza that are notable for the paucity of records of epidemics. These lean years have sometimes lasted for several decades. In 1940, Dr. Macfarlane Burnet,[1] of the Walter and Eliza Hall Institute in Melbourne, Australia, wrote:

> From 1848 to 1889 there was a rather extraordinary absence of influenza from England. A small number of deaths each year were registered as due to influenza, but no epidemics are on record, and it is reasonably certain that the virus was absent from England for 40 years.

He mentions four influenza epidemics recorded from continental Europe during that period "which failed to reach England," but his conclusion that the absence of records indicates absence of the virus cannot be accepted as "reasonably certain."

The lull in recorded epidemics in Great Britain was terminated by the pandemic of 1889–90, and the three epidemics that succeeded it, peaking in May 1891, January 1892, and December 1893. These four are often described as separate waves of a single pandemic but this may be seriously misleading. We have no reason to suppose that all were caused by the same influenza virus.

After 1893 came another lull in recorded epidemics until the arrival in 1918 of the most formidable influenza pandemic in human history. During the interval, however, the virus was not absent. Burnet considered that it had always been present with exacerbations in 1895, 1900, and 1908.

Earlier, the American epidemiologist, Wade Hampton Frost,[2] had provided evidence of its annual presence throughout that period. In 1919, long before the discovery of human influenza virus, he used the information supplied by excesses in morbidity and mortality not only attributed to influenza but also in the group diagnosed as "pneumonia (all causes)." Frost wrote:

> During great epidemics there are abundant, if not exact, records of prevalence, and the resulting mortality can be determined with fair precision, even though a large proportion of the deaths

are classified under diagnoses other than influenza. In the intervals between epidemics influenza becomes inextricably confused with other respiratory diseases, having a general clinical resemblance but no definite etiological entity, so that the record of prevalence and even of mortality is virtually lost. The first requisites for epidemiological study, namely, clear differential diagnosis and systematic records of occurrences, are therefore lacking in influenza.

In the absence of these essential records, statistics of mortality from the group comprising influenza and all forms of pneumonia afford, perhaps, the nearest approximation to a record of influenza. (pp. 321–325)

Frost warns that mortality from this combined group of diseases gives *no measure of the prevalence* of influenza, but it does provide an *index of its presence.*

. . . since it is well established that the epidemic prevalence of influenza markedly affects the mortality from this group of diseases, and since it is at least probable that even in nonepidemic periods there may be some intimate and constant relation between the prevalence of influenza and the mortality from pneumonia.

Frost and his colleagues based a study of influenza from 1887 to 1916 on this valuable principle. The state of Massachusetts possesses a record of deaths and of death rates per 100,000 of its population attributable to influenza and to all forms of pneumonia. Figure 3.1 depicts the findings from the Massachusetts archive. Epidemic influenza appears as an annual winter visitor.

Subsequent studies have validated Frost's method and they prompt the suspicion that influenza may have been behaving similarly in the British Isles from 1848 to 1889, and, indeed, that it may be an annual visitor to all large populations even though in some years its presence may be small. The theoretical importance of this finding will become apparent when we examine the explanations that have been offered of the behavior of epidemic influenza.

PRELUDE TO THE 1918 PANDEMIC

In December 1915 an abrupt rise occurred in the general mortality in many parts of the United States and by January 1916 an influenza epidemic had been reported in 22 of the states. The disease was not uncommonly severe and occasioned little public comment.

In Figure 3.2. Frost has used pneumonia and influenza mortalities to show the impact of epidemic influenza on three large cities. In all of them an unusually high pneumonia plus influenza mortality peaked in the first months of 1918, and Cleveland, Ohio, and New York City suffered high peaks in the two preceding winters. Frost[2] comments:

The rise in mortality from this group of etiologically heterogeneous diseases in the spring of 1918 is so sudden, so marked, and so general throughout the United States as to point very clearly to the operation of a single definite and specific cause, something largely independent

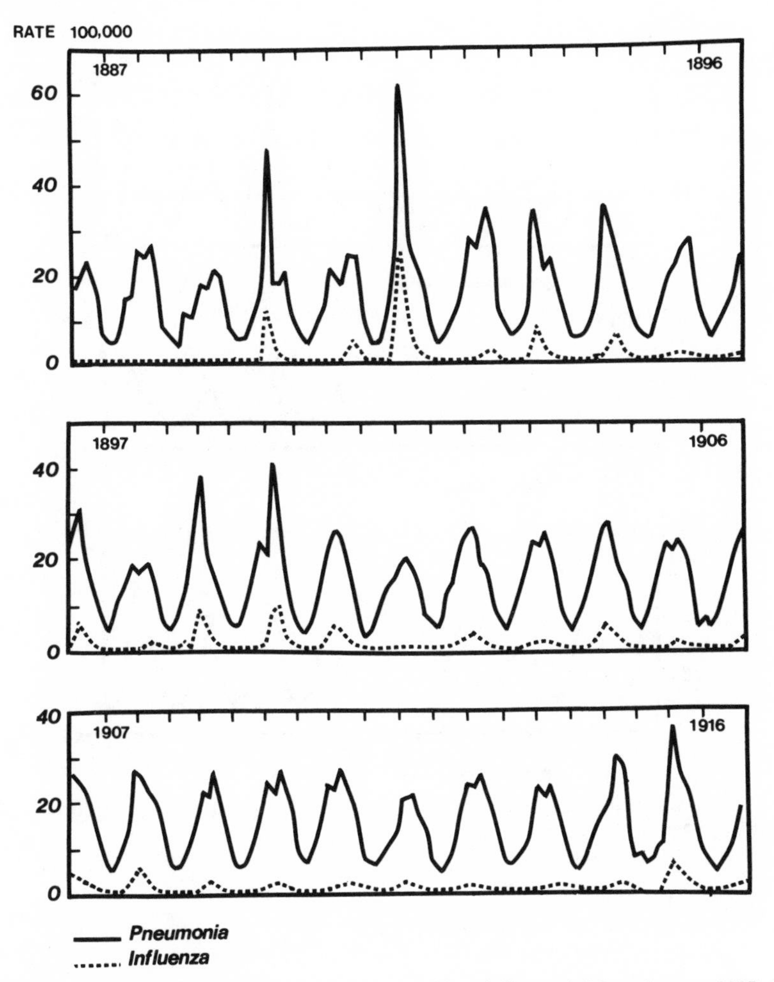

FIGURE 3.1. Mortality from pneumonia, all forms, and from influenza in Massachusetts, 1887–1916 (from Maxey,[2] p. 324, Chart 1).

> of meteorologic and other conditions. . . . the increased pneumonia mortality of March and April, 1918, was the consequence of a beginning and largely unnoticed epidemic of influenza, the beginning in this country of the great pandemic which developed in the autumn. (p. 329)

Europe also suffered mild influenza in the spring of 1918 and places as far apart as Great Britain, continental Europe, China, India, the Philippine Islands, and Brazil seem to have been contemporaneously attacked in June and July 1918, an unusual time of year for the communities living in the northern zones.

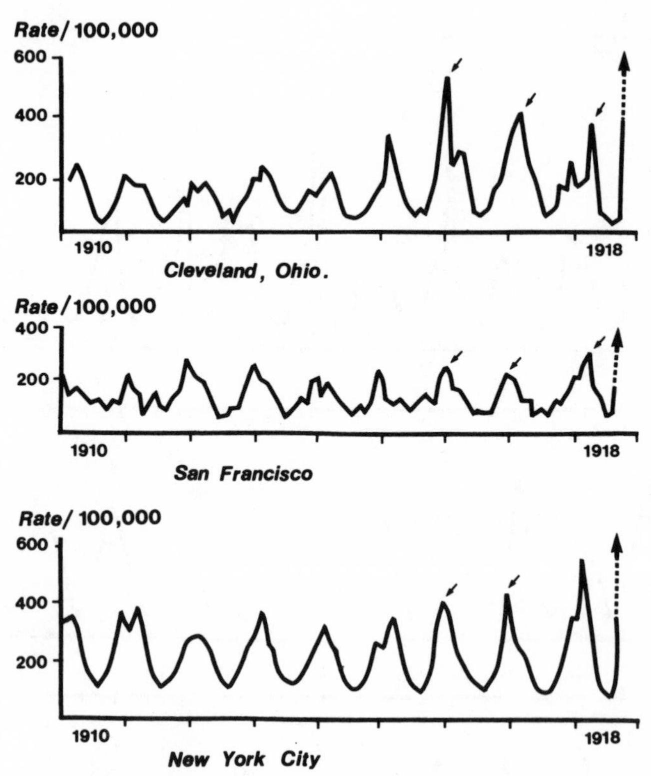

FIGURE 3.2. Death rate from pneumonia, all forms, plus influenza in three U.S. cities (from Maxcy,[2] p. 325, chart 2).

THE PANDEMIC OF 1918

I first became aware of the unique character of the disruption that influenza can cause when in October 1918, as a lad of ten years old, I was one of 90 boarders in a school near London attacked by so-called "Spanish flu." Only one of the boys escaped the disease and only one boy died of it. The school staff were so heavily attacked that parents of the schoolboys had to take over the school and nurse the sick. Much later I became aware that I was but one among more than a billion of the world's population to have been attacked, of whom some 20 million were said to have been killed by it.

It has become customary to describe the 1918 pandemic as having arrived in three waves, the first April–July 1918, the second October–November 1918, and the third February–March 1919. Three such waves of increased mortality coincide with records of influenza like epidemics in many areas and it has seemed reasonable to attribute all three to the activity of a single agent. Caution must, however, be exercised. Evidence is now available that at least three and possibly four influenza viruses may have been causing influenza during the period 1917 to 1919, namely two or possibly three influenza A viruses and influenza B virus. The evidence of their presence will be discussed in a later chapter. The epidemic in the spring of 1918 was not necessarily caused by the agent that produced the devastating pandemic in the autumn of that year. The so-called "third wave" in the early months of 1919 may well have been caused by the autumn virus, but here again there can be no certainty.

It was the so-called "second wave" that caused the shocking morbidity and mortality in most parts of the globe during the short space of time between late September and the end of November 1918, a phenomenon possibly unique in human history. The extensive literature to which it has given rise cannot be reviewed here, but no discussion of influenzal epidemiology is complete that omits to take note of the unusual behavior of the disease in the autumn of 1918.

The usual seasonal pattern of epidemic influenza in the world population over the surface of the globe will be described later because it is one of the aspects of the behavior of influenza that calls for explanation by any concept of its epidemiology, including the new concept advanced in this book. It would therefore be improper to omit the inconvenient fact that the autumn pandemic of 1918 behaved in a quite different manner in that it erupted contemporaneously in the northern and southern hemispheres. Such anomalous behavior must not be allowed to pass without comment, and a possible explanation will be discussed later. It is sufficient at this point to draw attention to the remarkable simultaneity of peaks of mortality and morbidity in widely separated areas. The peak of mortality in Cirencester, Gloucestershire differed from that in Auckland, New Zealand in the Antipodes some 10,000 miles away, no more widely than from that in Carlisle, Cumbria (Fig. 3.3A).

Figures 3.3B, 3.3C, and 3.4. compare the curve of the general mortality in 1918–19 in cities in the United States and Great Britain. Frost was an early exponent of using the general mortality as an index of influenza. A remarkable feature in 1918 was the high mortality among young adults, especially males.

1919–1928

Influenza epidemics decreased in severity as they came every year or two during the next decade. They caused the greatest mortality among the aged and persons already ailing with chronic bronchitis, emphysema, pulmonary tuberculosis, and heart disease. Small infants were also vulnerable. This age distribution of influenzal mortality marked a return to the earlier pattern, in contrast to that of the 1918 pandemic that preferentially killed young adults, especially heathy men.

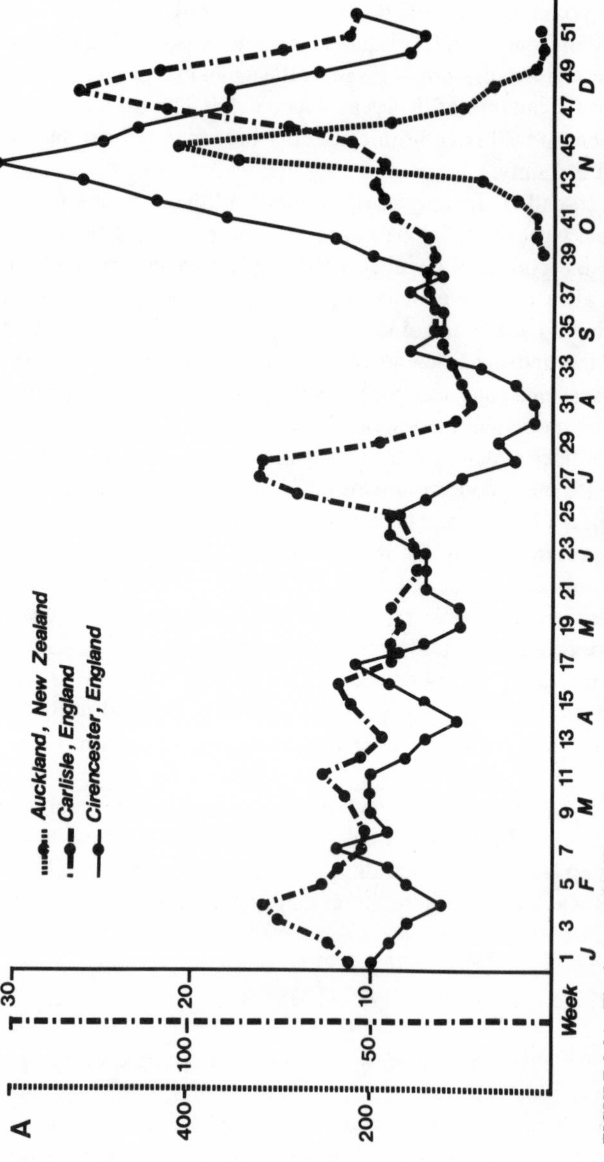

FIGURE 3.3A. The impact of influenza in autumn 1918 on mortality in: New Zealand (Bryder L., MA thesis, Oxford University, p. 21), Carlisle, Cumbria, and Cirencester, Gloucestershire, England. The synchronicity between New Zealand and the United Kingdom contradicts the usual global pattern of epidemic influenza.

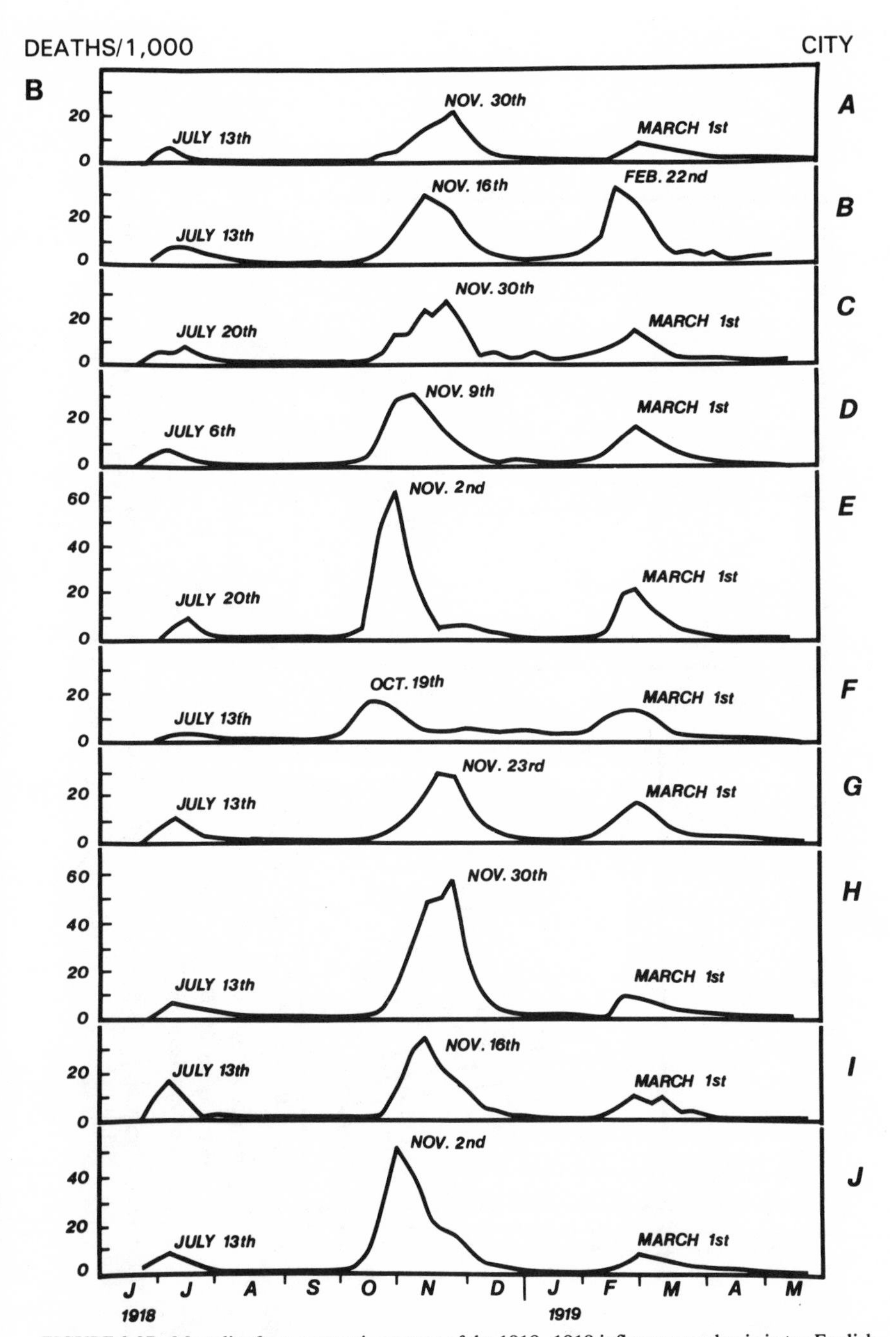

FIGURE 3.3B. Mortality from successive waves of the 1918–1919 influenza pandemic in ten English cities: A: Birmingham; B: Bradford; C: Halifax; D: Leeds; E: Leicester; F: Liverpool; G: Manchester; H: Nottingham; I: Oldham; J: Sheffield (from Maxcy,[2] p. 376, Fig. 4.4).

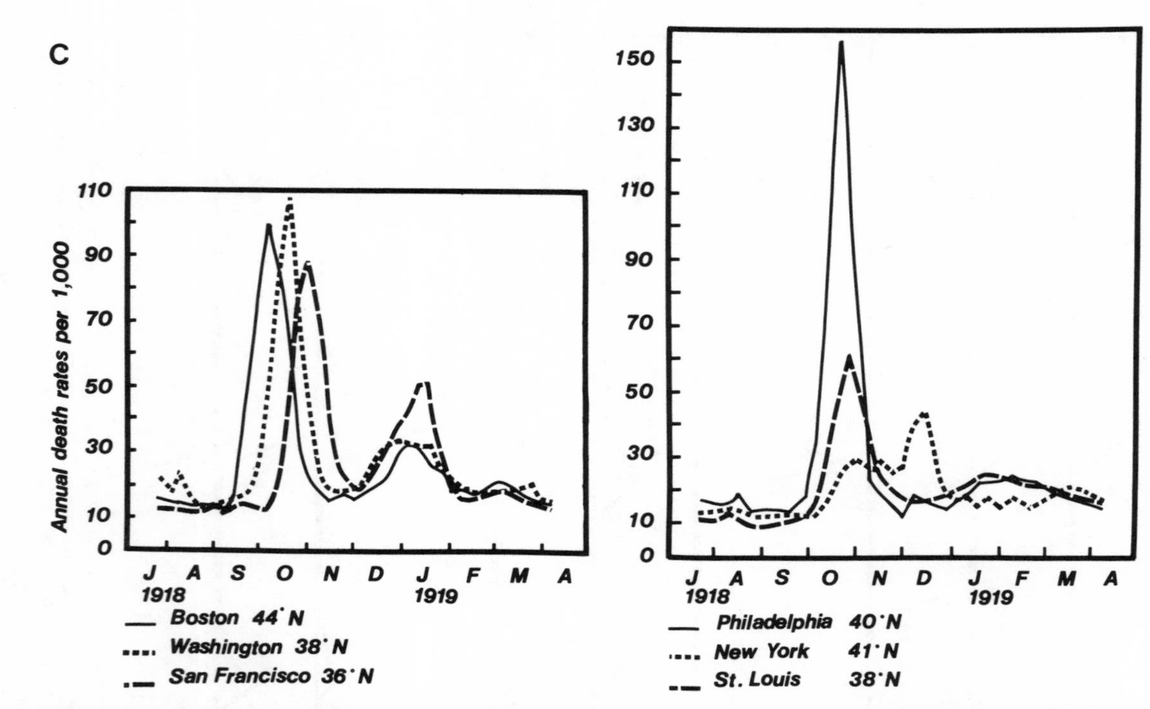

FIGURE 3.3C. Death rate per 1000 population from all causes in six U.S. cities, July 1918 to April 1919 (from Maxcy,[2] p. 332, chart 4, and p. 333, chart 5).

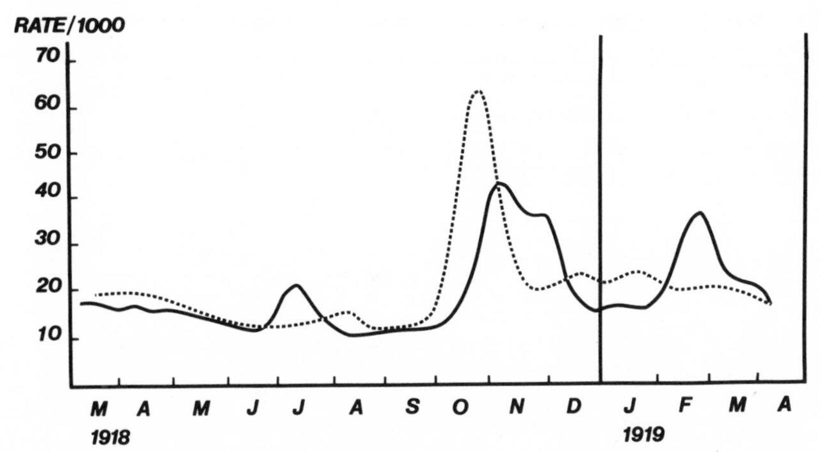

FIGURE 3.4. The impact of the 1918 influenza pandemic on death from all causes in 45 U.S. cities (. . .) and in 96 great towns in England and Wales(—). The peak mortality was nearly contemporaneous on both sides of the Atlantic Ocean (from Maxcy,[2] p. 330).

The preferentially high mortality of the 1918 autumn influenza may not have been a universal feature because Dr. Peter Gill of Beecroft, Sydney has been unable to confirm it from the Australian mortality data.

1928–1933

> An epidemic of such severity as that which the United States experienced in the autumn and winter of 1928–29 is, in its mass effect, among the most objective and impressive phenomena in the field of epidemiology. Over a wide expanse of country, within the space of a few weeks, the prevalence of acute illness is so greatly increased that it becomes a matter of common knowledge, apparent to everyone within his own circle of experience. The effect on mortality is equally or even more impressive, the *deaths from all causes* being sharply increased to one and a half, two or three times the usual number. (p. 427)

So W. H. Frost and V. A. van Volkenburgh began an article first published in the *American Journal of Hygiene* in 1935.[2] They had been studying the respiratory diseases of 1928–29, the year that experienced a great influenza epidemic, and comparing them with the following year, 1929–30, which was relatively free of influenza. The influenza epidemic in Baltimore lasted from 25 November 1928 to 23 February 1929.

It was a nationwide epidemic with an effect on the general mortality that shows it to have been the most severe since 1920. It had been severe enough in Baltimore to double the expected death rate from influenza plus pneumonia during three successive weeks, in one of which it had been tripled, and Baltimore was one of the less severely affected cities.

Subsequent serological studies suggest that the 1928–29 influenza epidemic marked the impact of a major antigenic change in influenza A virus, initiating an era of world prevalence of the novel strain that four years later was discovered as the first human influenza virus.

REFERENCES

1. Burnet FM: *Biological Aspects of Infectious Disease.* Cambridge University Press, 1940, p 244.
2. Maxcy KP (ed): *Papers of Wade Hampton Frost, MD.* New York: The Commonwealth Fund, 1941, pp 321–325, 329, 427.

4

The Effect of the Discovery of the Causal Organism

THE DISCOVERY OF AN INFLUENZA VIRUS

Doubts about the hemophilic bacillus that Pfeiffer had claimed in 1892 to be the cause of influenza had begun to develop as early as the turn of the century. Suspicion was already beginning to be focused on ultramicroscopic filter passing agents, some of which at that time had been identified as causing foot and mouth disease in cattle, human yellow fever, and a number of diseases in plants. One of these agents had been found to be causing epidemics in domestic poultry, but the avian illness was not then recognized as influenzal and the attribution had to await many years until other influenza viruses had been discovered.

Pfeiffer's *Haemophilus influenzae* was commonly isolated from persons suffering from influenzal pneumonia during the autumn pandemic of 1918 but seems to have been absent from the so-called first wave in the previous spring and summer. At the height of the pandemic attempts were made to transmit the causal agent of that terrible influenza by administering nasal washings taken from influenza patients in the early stage of the disease to courageous human volunteers and to nonhuman primates. The washings were first filtered through unglazed porcelain to remove all bacteria. These experiments were only partially successful as were those using unfiltered material. Pfeiffer's bacillus again became scarce in the influenza epidemics in subsequent seasons and a general opinion developed that the specific cause of influenza had yet to be discovered. The hunt turned in the direction of a filter-passing agent.

A major difficulty needed to be overcome. The filter-passing agents needed to be grown in living cells and could not be cultured on the media used for isolating and cultivating bacteria. Cell cultivation was in its infancy, and it was difficult to find an appropriate laboratory mammal as an alternative.

Since 1918 the domestic pig farms in the midwest of the United States had suffered from a seasonal epizootic disease much resembling human influenza, and in 1930 Richard Shope isolated the causal agent, the swine influenza virus.[1] To Shope must go the credit of identifying the first influenza virus. Despite careful epizootiological studies, Shope was unable to explain the behavior of the disease in herds of swine by simple direct case-to-case transmission of the virus. The seasonal epizootics arrived usually around October in what the local farmers described as "hog-'flu weather." Shope investigated the problem and his results led him to propose a complex solution that we shall be discussing in connection with other alternative concepts of influenzal epidemiology (Chapter 12). He repeatedly described how swine influenza breaks out simultaneously in widely separated farms with no intercommunication, suggesting that the virus must already have been seeded in the herds and activated perhaps by a meteorological stimulus. This led him to speculate that a similar situation might explain human epidemics of influenza.

THE DISCOVERY OF A HUMAN INFLUENZA VIRUS

Shope's discovery that a filter passer was the cause of swine influenza expedited the search for a similar organism in human influenza. The pig might have seemed the most promising alternative host in which to attempt to isolate the human influenza virus except that pigs are expensive, inconvenient laboratory mammals and already possess their own brand of the virus.

As so often happens in research, a fortunate accident facilitated the discovery of the human influenza virus. Work on various animal diseases had been in progress for many years in London at the National Institute for Medical Research in Hampstead. In 1926, Laidlaw and Dunkin[2] had discovered the virus that caused canine distemper, an epizootic disease that had been killing 50% of the puppies that were attacked and causing severe losses to the fox farms proliferating on both sides of the Atlantic. The Hampstead workers, in their attempts to produce a protective vaccine against canine distemper, accidently infected a ferret and so discovered that ferrets and other species of *Mustelidae* are even more susceptible than dogs to canine distemper virus.

When Laidlaw and his colleagues turned their attention to the identity of the agent causing the severe human influenza of 1932–33, they were unable to infect their laboratory mice. Not surprisingly they then attempted isolation in the ferret, which proved to be susceptible, and in 1933, Wilson Smith, Christopher Andrewes, and Patrick Laidlaw were able to announce the first isolation of a human strain of influenza virus.[3]

Accidents continued to hinder and to help. Andrewes had himself contracted influenza and Smith had collected a throat washing from him and had put it into a variety of experimental animals while Andrewes had gone home to bed.[4]

> . . . when I came back a few days later, the first ferret was sneezing. But then we ran into trouble because our stock of normal ferrets had got infected with [canine] distemper and for some little while we did not know where we were with a mixture of 'flu and distemper in the animals.[4]

Early in March, when the human epidemic was over, an infected ferret sneezed in Wilson Smith's face and he developed influenza, thus "cleaning-up" the virus by passing it through a human. Humans are not susceptible to canine distemper. The Hampstead team also found that the human influenza virus would infect mice if it had first been passed in a ferret.

A few years later Magill[5] and Francis[6] at the Rockefeller Institute in New York isolated a different virus from a mild influenza epidemic then current. There was no mutual cross-protection with Laidlaw's virus, so two types of influenza virus were evidently able to cause human epidemic influenza, and the Hampstead strain was named type A, the New York strain type B.

THE INFLUENZA OF 1932–33 IN A GENERAL PRACTICE

The epidemic in which the Hampstead workers first isolated influenza virus was widespread and severe. My first appointment in general practice began in September 1932 in a rural area of the County of Dorset. The work being quiet I obtained leave from December 21 to 28 for my marriage and honeymoon. When I returned to work a few days after Christmas, the transformation of the work in the practice came as a shock. Requests for visits poured in continuously, beginning before breakfast. Country rounds were so protracted that my wife did not know when or whether I should return for a midday meal. One day I found that I had traveled 88 miles and "treated" 176 patients. When summoned to a small village for a single patient, I found notes affixed to cottage doors requesting me to enter because the whole family was abed with influenza, and so I passed from house to house.

Despite large numbers of appropriate medicines carried in my car, my wife and I spent much time each evening preparing, wrapping, and addressing still more packages of medicine and taking them to the local powdered milk factory for dispatch at 5 AM next morning on the trucks collecting milk from farms in the surrounding countryside.

An almost intolerable pressure of work, with one or more calls each night, lasted for several weeks, leaving an aftermath of debilitated patients, some bereavements, and exhausted general practitioners.

Some of the earliest cases occurred in remote farms and cottages with little communication. The explosive onset over a wide country area was difficult to reconcile with the accepted picture of a disease agent being directly transmitted by the sick patient. Laidlaw, to whom I confided my doubts, urged that no sound epidemiological understanding could be obtained unless the name "influenza" were strictly confined to diseases caused by the newly discovered virus. The term was being loosely used by physicians and public as a cloak for ignorance of the etiology of an illness, much as the diagnosis "a viral disease" is often used now. Time has validated Laidlaw's dictum.

THE CONCEPTUAL IMPACT OF DISCOVERY OF THE INFLUENZA VIRUSES

The discovery of the virus initiated a spate of information about the causal organisms and the reactions to them in the body of the host. Contrary to expectation, far from solving the epidemiological difficulties, more problems demanding an explanation were coming to light year after year. Already by 1942 Andrewes was questioning whether the virus could be spreading directly from the sick:

> Where is the virus between epidemics? No carriers are found. . . . The virus disappears for 21 months out of 24. The British two-yearly rhythms seem to coincide with those on most of the European continent and in North America. So that one can hardly imagine that the virus keeps going by means of infection spreading from one place to another and finally going back to its starting-place. Not even if we bring in the Southern hemisphere can that theory be made to work.[7]

Shope, as we shall see, had postulated and produced some evidence for a cycle involving the lungworm and the earthworm to explain the epizootiology of swine influenza. Andrewes, whose hobby was natural history, was keenly interested in Shope's work and speculations: "It seems very likely that human influenza virus also can exist in occult form . . . not necessarily outside the human body."

Andrewes was picturing a basic influenza virus, stripped of the properties by which we are able to recognize it, existing between epidemics hidden in some human or other site: "A moderate view would picture epidemics of influenza not as arising from a single source, nor yet from latent infections of ubiquitous distribution, but rather from a limited number of scattered foci." This bold speculation was made many years before it was known or considered likely that type A influenza viruses are able to persist in various modes of latency and persistent infection of other host species and in cell cultures. Even now, in 1989, latent forms of influenza virus have not yet been definitely identified in influenza parasitism of mankind.

A young Australian, Frank Macfarlane Burnet, working under Frank Kellaway in the Hall Institute at Melbourne, was brought over to England at the request of Henry Hallett Dale, director of the National Institute for Medical Research, to work with Laidlaw's team. Burnet suggested to Andrewes that some of the problems of influenza might have a serological explanation. The earliest cases during an epidemic might occur in persons with low antibody protection, but as the epidemic progressed, the virus might develop progressively greater potency so that it became able to breach the immunity of persons possessing more antibody: thus when virulence had been stepped up locally, further spread by droplet infection could occur in the orthodox way.

Something similar had been shown to occur during epidemic spread of certain bacterial pathogens. Nevertheless Andrewes felt it necessary even in these early speculations to postulate some mode of latency for the influenza virus, but he was reluctant to allow it as a widespread characteristic. He was still attempting to retain the major role for the measles-type of model of direct spread from the influenzal patient. The attempt was to be fraught with difficulty and to occasion many conceptual twists and turns.

Burnet also found himself compelled to adopt the view that the virus can persist in some noninfectious mode. In a lecture at Harvard University in 1944 he said: " . . . as yet there is no visible alternative to the view that human influenza viruses survive between epidemic periods in the tissues of human carriers."[8]

He envisaged a system similar to but not identical with that of the cold sore virus *Herpesvirus hominis*, a system in which a group of respiratory tract cells harboring latent influenza virus would be reactivated to infectiousness by cold weather or some other particular climatic stimulus.

FIELD STUDIES THAT STIMULATED NEW IDEAS

Through the next decade the difficulties of explaining the behavior of epidemic influenza became increasingly apparent. In 1949, Professor F. Magrassi of the University of Sassari wrote an account of his careful study of the influenza epidemic that had attacked Sardinia in the autumn of 1948.[9] There were two main groups of foci in relation to the start of the epidemic. The first included ten villages in which influenza began simultaneously during mid-September and peaked a month later. The second group involved 16 villages where the epidemic appeared in the latter part of October or later, at the same time as it began on the mainland of Italy, near Rome.

Magrassi is confident that the disease arose independently and spontaneously in each of the inhabited centers of the first group because they are widely separated and no cases had been reported before they began. he claims to be the first person to report the multicentric origin of an influenza epidemic in such a precise manner.

He further maintains that some cases in the second group may also have originated spontaneously and not be spread from a neighboring source. He was particularly impressed by cases among shepherds living in open country in complete and long isolation who had developed influenza contemporaneously with inhabitants of towns and villages many miles away.

Magrassi discusses the problems in the light of the then-available knowledge of the virus and the host response to it and in relation to a number of hypotheses. He concludes:

> Changes such as those observed experimentally . . . may occur spontaneously in nature, when there are conditions favorable for the "transformation" of the virus. As for the multicentric origin of the epidemic in Sardinia, one might suggest that conditions favourable to a change developed because of interaction of two important factors: the host (man) carrying a latent infection and the environmental pressure inducing the change in the virus. Such an interaction cannot be fortuitous or exceptional as it was repeated simultaneously in many centers in the same environment.

Here we have an embryonic statement of the new concept advanced in this book, except that he assumes that the carrier develops influenza at reactivation and then transmits the virus from his illness. Later we present evidence against such timing (Chapter 7).

Observers on both sides of the Atlantic and in the Soviet Union and Australasia were discussing various concepts of the epidemic mechanisms and Andrewes was at the forefront of many such debates. He warned against thinking in terms of our English winters when discussing the seasonal nature of influenza, because we must bear in mind that influenza is ubiquitous, occurring in places where winter as a cold season does not exist. But the trap against which he warned is insidious, and later we find him writing: "This dependency of influenza on season must engage our careful attention. . . . I shall refer to the strange operative effect of season as 'winter factor.'"[10]

One must, however, be grateful that he was attempting to focus attention on the need to explain the seasonal nature of the disease. This is a salient feature of the new concept.

The National Institute for Medical Research at Hampstead was then one of the few centers collating reports of influenza and receiving specimens of the virus from many parts of the world, and the workers there were intrigued by the timing of epidemics, sequential or contemporaneous, in different countries. In 1949 Andrewes was questioning whether, as generally supposed, the influenza virus does really spread from country to country, or alternatively " . . . *whether circumstances change in series from one country to another, activating endemic virus in each and creating an illusion of the spread of the infection.*"[11]

I have italicized his speculation because it too anticipates a key proposition of the new concept discussed later in this book. Unfortunately, Andrewes promptly retracted his speculation:

> Fortunately the studies made this year [1949] seem conclusive that a genuine spread has occurred because the 1949 viruses from Italy, France, Switzerland, Great Britain and Iceland were of one antigenic type close to but distinct from A prime strains obtained in Australia in

> 1946 and in Europe and America in 1947. Amongst themselves the 1949 viruses were of remarkable homogeneity.

This assumption, that the finding that viruses isolated in different countries are identical proves that they have been traveling by direct spread, has been a stumbling block. The assumption is dubious because the picture is equally well explained by a concept of virus latency with seasonal reactivation as proposed later in this book. Subsequently, we shall discuss the epidemiological features that are required to prove that direct spread is occurring in infectious diseases and also other important features of the host–parasite relationship.

Doubt about the adequacy of the concept of direct spread to explain how influenza was actually behaving soon returned. In 1951, Andrewes wrote:

> I have previously suggested that between epidemics flu virus may exist in a modified phase. . . . There is no certainty as to whether influenza really spreads from country to country as it seems to do, or whether endemic viruses are successively activated in different countries producing the illusion of invasion across frontiers. . . . Unless all our thoughts are wrong *an influenza epidemic must emerge where there is an underground virus to emerge*. . . . A point to note in the beginnings of epidemics is their apparent multifocal origin.[12]

Andrewes was familiar with Magrassi's observations and had been much impressed by his findings.

TRANSEQUATORIAL SWING OF INFLUENZA

Observations in South Africa and southern Australia drew attention to a global phenomenon that was to tax still further the imagination of epidemiologists. Epidemics of influenza in these southern latitudes regularly preceded or succeeded those in northern latitudes by about six months. How could the virus be achieving such a feat by a mechanism of direct spread? Burnet proposed a hypothesis of transequatorial swing whereby influenza virus might be able to keep traveling, presumably from patient to patient, continuously around the world population, always in the winter months. He made no estimate of the speed that it would need to be traveling in order to cover such vast distances regularly within the specified time. Andrewes, the keen natural historian was fascinated by the problem:

> Can it be that 'flu can only keep going in the winter to and fro across the equator, much as the Arctic tern migrates yearly from the Arctic to the Antarctic and back? . . . The epidemics could have origins of two sorts, one from beyond the equator, another from nearer home.[10]

Many years later it was found that influenza A viruses are in fact harbored by terns and other seabirds. The new concept, however, proposes a simpler explanation of transequatorial swing.

THE ROLE OF HERD IMMUNITY

Soon after discovery of the virus in man the early workers on influenza became aware that "human herd immunity" to the disease, the immune response to infection of previously infected individuals in the community under attack, must be a major factor in determining the epidemic behavior of the causal parasites. Thus in 1951 Andrewes writes:

> It would seem possible that emergence [of the virus from its postulated underground condition] would be hindered by a high level of herd immunity. At the fringe of its exploits, however, herd immunity would be lower . . . and its reappearance might become more possible.[12]

It will appear later that human herd immunity does indeed play a most important role in the epidemic mechanisms, but lack of precise definition of the term has sometimes led to confused epidemiological speculation so that the concept of herd immunity has been wrongly invoked to explain situations in which chiefly nonimmune persons are being infected.

A NEW HYPOTHESIS FROM ANDREWES

In 1952 Andrewes introduced a new conception that increased the credibility of his later speculations.[13] He had shown experimentally that he could produce at will either an attack of influenza or a subclinical inapparent infection in the animals he infected with influenza virus by adjusting the infecting dose of the virus. He reasoned that in the human situation:

> It may be that . . . if you . . . encounter first a small dose of the virus, your basic immunity is stimulated to permit rapid overcoming of a bigger dose later on. If you are less lucky, you meet a heavy dose initially and then go down with flu.

This attractive suggestion also provided a plausible explanation for the phenomenon of antigenic drift of the influenza virus that we shall be discussing later (Chapter 13).

THE VANISHING TRICK

Information and specimens reaching Hampstead from many parts of the world drew attention to the most puzzling and apparently illogical of the many conundrums posed by the human influenza viruses. The strains discovered in 1932–33, soon to be called type A influenza virus, remained homogeneous for more than a decade. Then in the winter of 1946–47 they were replaced by a different but related virus that was named "A prime" (written A') in order to distinguish it from the earlier A strains (See Table 5.1 in Chapter 5, p. 48). Vaccination by a vaccine containing the original A virus conferred little protection against the novel strains. Andrewes was deeply puzzled: " . . . strange as it may seem these A primes seem to have completely replaced the classical As all over the world. How this comes about and why the classical As should have vanished is a mystery."[13]

A mystery indeed! The phenomenon, christened the vanishing trick, has characterized most subsequent major and minor antigenic changes of influenza A virus. Strains that have been causing all the type A influenza in the world for perhaps a dozen years will vanish and next season be replaced everywhere by a novel strain. In the case of minor antigenic changes the predecessor may have been prevalent for only one or two seasons over a large part of the earth's surface before it disappears and is replaced by a new minor variant.

The vanishing trick still remains to be explained. In 1975, Edwin Kilbourne wrote:

> . . . no less remarkable than the sudden appearance of major antigenic variants of influenza A virus as a concomitant of pandemic disease is the seemingly simultaneous disappearance of the antecedent virus from natural circulation. This is all the more remarkable because of the retention and preservation of potentially infective influenza A subtypes of the past in virological and diagnostic laboratories.[14]

Robert Webster and Graeme Laver after a discussion of the problem in 1975 concluded that there is no satisfactory explanation of the phenomenon available.[13]

A phenomenon that seems to defy logical explanation is surely calling attention to some fundamental feature of the epidemic mechanism and poses a challenge that cannot be ignored by any valid concept of the epidemiology of type A influenza. It is not yet certain if the vanishing trick also characterizes the behavior of influenza B viruses.

Already in this chapter it has become apparent that the development of hypotheses about mechanisms operating influenzal epidemicity is depending increasingly on rapid advances in knowledge of the structure of the virus, the functions of its different parts, and its relationship with the host cell and the immune systems of its human host. In order that readers may appreciate the way in which new concepts have developed and may understand the debates that have arisen, the next chapter provides a simple account of the virus and its method of replication, and of the way in which its structures relate to its behavior.

REFERENCES

1. Shope RE: Swine influenza. III. Filtration experiments and etiology. *J Exp Med* 54:373–385, 1931.
2. Laidlaw PP, Dunkin GW: Studies in dog distemper. III. the nature of the virus. *J Comp Pathol Ther* 39:222, 1926.
3. Smith W, Andrewes CH, Laidlaw PP: A virus obtained from influenza patients. *Lancet* 2:66, 1933.

4. Andrewes CH: Influenza A in ferrets, mice, and pigs, Stuart-Harris CH, Potter CW (in eds): *The Molecular Virology and Epidemiology of Influenza*. London; Academic Press, 1984, pp 1–3.
5. Magill TP: Virus from cases of influenza-like upper respiratory tract infection. *Proc Soc Exp Biol Med* 45:162, 1940.
6. Francis T: New type of virus from epidemic influenza. *Science* 92:405, 1940.
7. Andrewes CH: Thoughts on the origin of influenza epidemics. *Proc R Soc Med* 36:1–20, 1942.
8. Burnet FM: *Virus as Organism*. (The Edward K. Dunham lectures for 1944). Cambridge, Mass, Harvard University Press, 1945, p 105.
9. Magrassi F: Studies of the influenza epidemics in the autumn of 1948. *Minerva Med* 1(19):565–569, 1949. (In Italian. English translation thanks to Professor Negroni).
10. Andrewes CH: Factors in virus evolution. *Adv Virus Res* 4:1–24, 1957.
11. Andrewes CH: Recent advances in knowledge of influenza. *Proceedings of the Annual Meeting of the British Medical Association: Section of Preventive Medicine*. London, Butterworth, 1949, pp 171–176.
12. Andrewes CH: The epidemiology of influenza in the light of the 1951 outbreak. *Proc R Soc Med* 44:803–894, 1951.
13. Andrewes, CH: Prospects for the prevention of influenza (James M. Anders Lecture xxvii). *Trans Stud Coll Physicians Phila* 20:1–8, 1952.
14. Kilbourne ED: Epidemiology of Influenza in Kilbourne E D (ed.): *The Influenza Viruses and Influenza*. New York, Academic Press, 1975, pp 496–497.
15. Webster RG, Laver WG: Antigenic Variation of Influenza Viruses, in Kilbourne ED (ed). *The Influenza Viruses and and Influenza*, New York, Academic Press, 1975, p 309.

5

The Viruses that Cause Epidemic Influenza

INTRODUCTION

If anyone wishes to understand the epidemiological problems presented by the influenza virus and the attempts that have been made to explain them, it is necessary to know something of its structure and physiology and of the appropriate terminology. The following brief description should suffice for the purpose of this book, and readers who are interested to go more deeply into the rather complex subject should consult a text book such as Kilbourne.[1]

There are at present three known types of influenza virus, namely types A, B, and C. Only viruses belonging to types A and B cause influenza epidemics, so type C will receive little mention in this book. Type A influenza viruses are widely distributed among many vertebrates besides mankind, and they present more problems and have been better studied than type B influenza viruses. They therefore receive more attention..

THE INSTABILITY OF RNA GENOMES

All nucleated organisms and many bacteria and viruses base their reproduction on a DNA genome, whereas influenza viruses are among a minority of organisms that base their replication on RNA. The properties of RNA help to explain the epidemiology of influenza and the difficulty of achieving successful prophylaxis by means of vaccines.

DNA and RNA genomes both use polymerase enzymes for replication, but under suitable conditions, such as might have existed in the primal metabolic soup wherein life is supposed to have originated, RNA molecules can replicate spontaneously and maintain spontaneous synthesis. An RNA gene may therefore have been the chemical progenitor of all living organisms both RNA and DNA based. The known RNA organisms are, however, all parasitic on DNA organisms or in some other manner dependent on them.

Mutations result from errors in the fidelity of replication of RNA or DNA genomes. Mutational frequency provides an estimate of the instability of these molecules. DNA, the more stable, is the more widely distributed, and DNA genomes have evolved mechanisms for reading information and for repairing errors in order to produce viable genome copies. RNA genomes, being smaller and simpler, consume less energy.[2]

The two sorts of genome differ much in evolutionary plasticity. Continuously replicating RNA viruses may evolve a million times more rapidly than their DNA-based host. The mutation rate is inversely proportional to the size of the genome.[3]

Because the error rate of RNA is so high, many variants of an RNA virus may coexist and compete. The fittest of these variants—perhaps the most stable, fecund, and best adapted to its host and the complexities of transmission—becomes the most abundant. We shall encounter this situation frequently in the behavior of influenza viruses. It has been described as a *quasispecies distribution* in which a *master sequence* is surrounded by a swarm of mutants from which it may be unable to escape. The situation sometimes stabilizes and retards the evolution of the genome by prolonging the dominance of the master sequence.[4]

Mutability is a powerful influence driving evolution, but it is not the only one (see Chapter 5: Antigenic Shift). The section of Chapter 12 on evolutionary dendrograms discusses recent contributions of molecular virology to the study of the evolution of influenza A virus.

THE STRUCTURE OF INFLUENZA A AND B VIRIONS

The type A influenza virion structure cannot be distinguished from that of type B. Figure 5.1 suffices for both. The virion adopts many shapes during infection with spheres, irregular rhomboids, and filiforms all being common. The pleomorphism may facilitate attachment of the parasite to a receptor area on the host cell membrane.

The membrane coating the virion is a lipid substance identical with that of the plasma membrane coating the respiratory tract cells of the host. The viral membrane is stiffened by an underlying shell of *matrix protein* (M). About 500 *spikes* projecting from the viral surface are rooted in the viral membrane. Some 400 of them are molecules of the glycoprotein *hemagglutinin* (H) by means of which the virus attaches itself to the host cell. The remaining 100 spikes, randomly scattered among the H spikes, are molecules of another glycoprotein, the enzyme

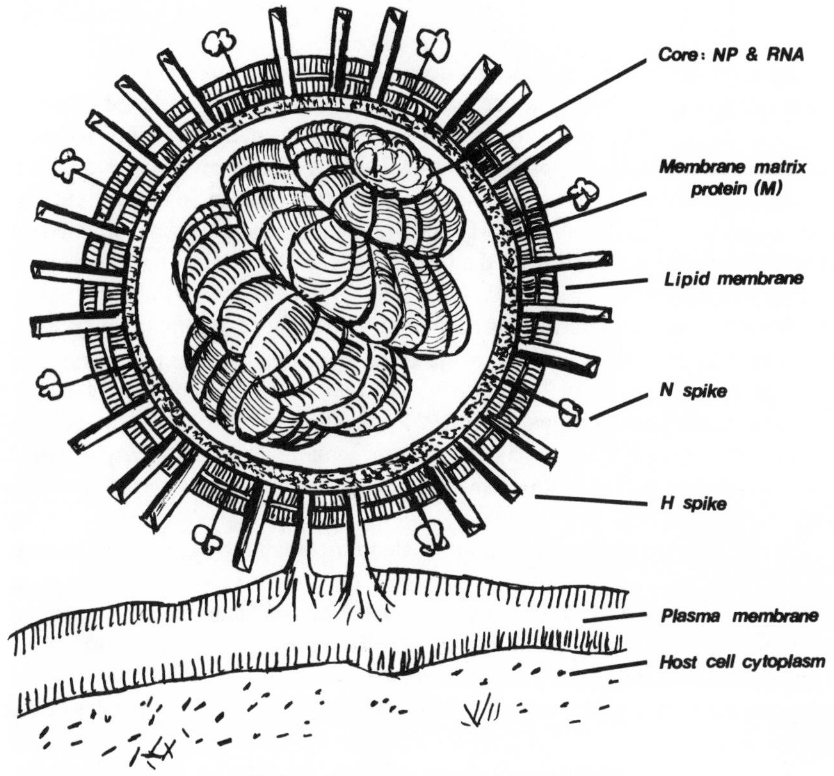

FIGURE 5.1. Diagram of an influenza virion.

neuraminidase (N), which probably acts on their sialic acid attachment to release newly formed virions from the cell in which they were produced. Though both surface proteins play a role in influenza epidemiology, H is the more important.

Within the cavity of the virion lies the core containing the single chromosome of the RNA genome in a scaffolding of *nucleoprotein* (NP). The structure is unusual. The eight RNA segments of the genome are each more or less encased by a separate rodlike scaffold of NP, each combination forming a *ribonucleoprotein* (RNP) *segment.* There are thus eight loosely associated RNP segments in the core structure.

It would be convenient if each segment contained a single RNA gene coding for a single viral protein as in segment 4, which harbors the gene coding for H, and in segment 6 containing the N-coding gene. There are, however, ten genes to be accomodated in the eight RNA pieces, so some contain more than one gene. Moreover, some genes have their message located in more than one segment.

CHAPTER 5

INVASION OF THE HOST CELL AND REPLICATION

As soon as the influenza virion encounters the respiratory epithelium of the nonimmune infected person, a receptor area on the surface of a ciliated cell or a mucus-secreting goblet cell recognizes receptor sites on contiguous viral H spikes. The virion is attached and promptly drawn into a coated pit that develops to receive it within the thickness of the cell membrane. There, in an acidic fluid, the H molecule is split into two parts, H1 and H2, by a protease enzyme supplied by the cell. No further penetration by the virion can take place unless the H spike has been split in this manner.

A tryptic enzyme in the lumen of the respiratory tract can sometimes split the H molecule before penetration into the cell.[5] Bacteria, for example, *Staphylococcus aureus*, can supply the enzyme, an association that may increase the severity of the influenzal illness, as in 1957 when the epidemic was characterized by numerous cases of staphylococcal pneumonia, many fatal. *Haemophilus influenzae* played an even more lethal role in the autumn influenza pandemic of 1918.

As the penetration progresses, the coated pit containing the virion is moved into the host cell's cytoplasm, becoming a *receptosome* in which the virion is uncoated. The core is then transported into the cell nucleus where replication begins at once with help from the genetic machinery of the cell. Much of the detail of the replicative process is now known, some of it taking place within the nucleus and the rest in various parts of the cytoplasm with the help of the organelles of the cell.

The viral RNA (vRNA) is single stranded and of negative sense. Within the nucleus it is copied to form messenger RNA (mRNA). The machinery whereby this is translated in the cytoplasm to complementary (cRNA) and finally to new vRNA is not yet well understood, nor is the manufacture of all the new viral proteins[1] (pp. 62–64).

Within a few hours of penetration, 100 or more incomplete virions are ranged at the periphery of the cytoplasm underlying the cell membrane. They become wrapped in their lipid coats as they are being extruded through the plasma membrane to emerge as little buds on the cell surface. There they remain until their N enzyme releases them to invade and destroy the next appropriate cell that they encounter.

Each invasion is lethal to the host cell, and the dead cells are shed into the respiratory passages. The resulting gaps in the epithelium are covered by the underlying stratified cells until within a few weeks the breaches are restored by new ciliated and goblet cells.

Deeper and more permanent damage may result when bacteria take advantage of the epithelial breaches to invade the walls of the bronchi, and they may reach the bronchioles and lung alveoli.

Within each influenza patient an astronomical multiplication of virions occurs. Each infected cell multiplies the virus 100- or 200-fold, and millions of cells are infected. The number of virus particles produced during the single season of a world influenza pandemic defies imagination, but this phenomenal reproduction should be borne in mind when considering the evolutionary potential of influenza viruses and the restraints that must operate to limit the number of variants, both major and minor, to the handful that we experience in practice.

THE ANTIGENICITY OF INFLUENZA VIRUSES

The proteins and some of the carbohydrates of the influenza virion provoke an immune reaction in the host that involves the production of local and humoral antibody and of cell-mediated immunity.

The nucleoprotein of the viral core provokes the antibody that is type specific, distinguishing between viruses of types A, B, and C.

Both surface proteins, the H and N spikes, possess several sites from which H and N antigens provoke antibodies that moderate and shorten influenzal illness and protect against reinfection by the same or by closely related influenza viruses. The antibodies, which are highly specific, appear in the circulation several days after the onset of illness and rapidly increase in amount for about a fortnight. They are used to identify the precise strain of the infecting virus, but, as they may persist for a lifetime, one must demonstrate the increase in antibody between an early specimen of serum and another taken a week or two later to identify precisely the virus causing that particular illness.

In common with other viruses, all three types of influenza virus are subject to variation known as *antigenic drift*. Influenza A viruses are also subject to a different sort of variation called *antigenic shift*. As currently defined, drift is caused by mutation in the genes coding for H and for N whereas shift is caused by genetic reassortment involving at least the gene coding for the H. Both sorts of antigenic variation are so important practically and theoretically that they receive extended consideration later in the book (Chapters 9 and 10).

CLASSIFICATION OF INFLUENZA VIRUSES

The warning given by Andrewes in the preface of the first edition of *Viruses of Vertebrates*[6] still applies in 1989, namely, that we do not know enough to classify all viruses in an orderly manner. Influenza viruses have been placed in the family Orthomyxoviridae and in the genus *Influenza virus*. The genus contains only the three types—A, B, and C—and the inclusion of type C is provisional.

All three types occur naturally in humans, and types B and C are primarily human parasites. Type A influenza viruses occur naturally in many other vertebrates besides man.

As related above, type A virus, unlike the other two, has subtypes based on the nature of the surface antigens hemagglutinin and neuraminidase. The three known subtypes are now classified as H1N1, H2N2, and H3N2, all of which have also been found in avian host species. Numerous other influenza A virus subtypes have been found in nonhuman species of host. At the time of writing there are 13 known H subtypes and 9 known N subtypes in more than 30 combinations, all of which have been found in waterfowl.

Reassortment of genes occurs so readily in many hosts that genes may travel between host species independently of the virus in which they originated by thus changing their packaging in viruses.

Table 5.1 shows how the classification of human influenza A viruses has been altering during the last 40 years as new discoveries have rendered previous classifications obsolete. These RNA-based viruses present peculiar taxonomic difficulties and the present classification may have been revised by the time this book is published. The reasons for each of the changes in nomenclature in Table 5.1 will be apparent in later chapters. Antigenic shift of type B influenza virus has not yet been found. Why it differs is not known.

TABLE 5.1. Successive Classifications of Human Influenza A virus[a]

(i) Date	(ii) 1946–57	(iii) 1957–68	(iv) 1968–79	(v) 1980–
1889–1900				[A(H2N2)]
1900–17				[A(H3N2)]
1907–17				[A(H1N1)]
1918–29			[A (Hsw1N1)-like]	[A(H1N1)]
1929–46	A	AO	A(H0N1)	A(H1N1)
1946–57	A'(A prime)	A1	A(H1N1)	A(H1N1)
1957–68	A" (A double Prime)	A2 Asian	A(H2N2)	A(H2N2)
1968–		A2 Hong Kong	A (H3N2)	A(H3N2)
1977–			A(H1N1)	A(H1N1)

[a]Note that in 1980 (column v) three major variants from column iv were combined, thus concealing that old style A(H1N1) in column iv returned in 1977 and is still prevalent. Moreover, strains of A(H1N1 old style) were also prevalent from 1907–1917. Square brackets show that the evidence was obtained from retrospective serology (from Hope-Simpson,[7] p. 104; reproduced with permission from *PHLS 7Microbiology Digest*).

STRAIN NOMENCLATURE OF INFLUENZA VIRUS

The designation of a strain gives the type, the place of its isolation, the number of the isolate, and the year of isolation as, for example, B/Hong Kong/1/79. For type A influenza strains it is necessary to add the subtype thus: A/Victoria/3/75 (H3N2). When the isolation has been made from a nonhuman host, the species of the host is inserted after the type: A/swine/Iowa/15/30 (H1N1).

Usually an early isolate is chosen as prototype for the minor variants. Thus most influenza A isolates in 1974–75 season were identical with the prototype strain: A/Port Chalmers/1/73 (H3N2), although the local laboratory identification might have been, for example: A/Leningrad/476/74 (H3N2).

For epidemiological purposes this prototype resemblance is most useful and the above (imaginary) Leningrad isolate would be called A/PC/1/73(H3N2)-like, but the word “like” is usually omitted.

REFERENCES

1. Kilboune ED: *Influenza*. Plenum Medical, New York, 1987.
2. Eigen M: Self-organization of matter and the evolution of macromolecules. *Naturwissenschafte* 58:465–523, 1971.
3. Eigen M, Gardiner W, Schuster P, Winkler-Oswatitsch R: The origin of genetic information. *Sci Am* 244:88-118, 1981.
4. Drake JW: Spontaneous mutation. *Nature*. 221:1128–1132, 1969.
5. Barbey-Morel CL, Oeltmann TN, Edwards, KM *et al*: Role of respiratory tract protease in infectivity of influenza A virus. *J Infect Dis* 155:667–672, 1987.
6. Andrewes CH; *Viruses of Vertebrates*. London; Balliere, Tindall Cox, 1964.
7. Hope-Simpson RE: Simple lessons from research in general practice. Part 8: The problems of antigenic shift. *PHLS Microbiol Dig* 7(4):104, 1990.

6

Antigenic Variation and Recycling of Influenza A Viruses

ERAS OF PREVALENCE OF A(H0N1) AND A(H1N1 OLD STYLE) INFLUENZA VIRUSES

Soon after discovering the first human influenza virus in 1933, the Hampstead workers noticed that the type A strains isolated in different parts of the world in a single season possessed a remarkable homogeneity, whereas those isolated even from the same area in successive seasons were distinguishable serologically, although vaccine prepared from isolates collected in any season provided protection against the strains from that and all other seasons. This must have been the first recognition of the occurrence of antigenic drift.

In 1946, however, they encountered an epidemic caused by an influenza A virus that had changed so much that vaccines prepared from strains isolated in earlier seasons were no longer protective. The earlier strains had disappeared as soon as the novel strains appeared in 1946, although serological studies showed that the earlier strains had hitherto been causing all the type A influenza in the world since 1929. As we saw in the last chapter, the new strains were for many years considered as belonging to a different subtype and the 1946 antigenic change was at first classified as a shift. Table 5.1 in Chapter 5 shows the names that were used to distinguish the two supposed subtypes and the series of name changes as the structure of the viruses became progressively clearer until by 1979 they were called A(H0N1) and A(H1N1), respectively. The A(H1N1) strains having replaced their predecessors worldwide in 1946–47 season continued to cause all the influenza A until replaced in 1957 by strains of another subtype, A(H2N2). In 1980 the two earlier eras and their causal viruses were amalgamated in a single subtype called A(H1N1) that included also the still earlier strain, A(Hsw1N1), serologically found to have caused the influenza A from 1918 until 1929.

Langmuir and Schoenbaum[1] in the United States listed the following epidemic seasons of a A(H0N1) influenza: 1932–33, 1936–37, 1938–39, 1940–41, and 1943–44. The British Isles experienced A(H0N1) influenza epidemics in each of those seasons.

During the era of world prevalence of A(H0N1) strains, workers were optimistic that a vaccine containing the then-current types A and B influenza viruses, if used widely enough, would prevent epidemics and might even stamp out influenza. The optimism was not unreasonable, the strains of both types being so homogeneous that North American type A isolates were identical with British type A isolates of the same season, and persons attacked in one season usually escaped in subsequent type A epidemics despite changes in the virus caused by antigenic drift. Vaccines prepared from 1933 A(H0N1) strains protected ferrets and mice challenged by the much drifted 1944 strains.

Hopes of an effective anti-influenza vaccine received a setback in the 1946–47 season when the epidemic of influenza A was found to be caused by a new strain against which neither a previous attack of A(H0N1) influenza nor the vaccine containing the strain afforded adequate protection. The immunology of influenza A viruses was not as straightforward as had been anticipated.

The antigenic change in 1946–47 was later found to have resulted from a mutation in the gene coding for hemagglutinin (H) similar to but larger than those that had been causing antigenic drift. It demonstrated that mutations on the H-coding gene may sometimes be sufficient to cause the changes that characterize antigenic shift; namely, not only the vanishing trick, which frequently occurs over smaller areas at seasonal antigenic drifts, but also a bypassing of the immunity conferred on persons previously attacked by the predessor strain and worldwide replacement by the novel virus in a single season. The mutant also initiated a family of successive minor mutants.

We speak of the replacement of A(H0N1) by A(H1N1 old style) strains as if the major mutation had taken place in the 1946–47 season. While such may have been the case, there is no evidence for it. The mutation may have occurred previously, perhaps decades or centuries before, and the virus or its genome may have been stored somewhere for recirculation when opportunity offered.

Langmuir and Schoenbaum listed the following A(H1N1 old style) epidemics in the United States: 1946–47, 1949–50, 1950–51, and 1952–53. Once again our experience in the British Isles was identical, and after their paper was written we and they both experienced epidemics caused by the same strains of A(H1N1 old style) in the 1954–55 and 1955–56 seasons.

The epidemic of 1950–51 was notable for its severity and its worldwide distribution. Two minor variants of A(H1N1 old style) virus known as "Scandinavian" and "Liverpool" were attacking populations in many parts of the world and they reappeared to cause another severe epidemic in the 1952–53 season. The Scandinavian strain has obtained permanent notoriety in influenzal history by returning in 1977 after 25 years absence to initiate a new era of prevalence of

A(H1N1) subtype. The Liverpool variant is also notable because it occasioned a higher influenzal mortality rate in 1951 in the city of Liverpool than that caused by the 1918 influenza pandemic. Had the 1950–51 epidemic occurred soon after the first appearance of A(H1N1 old style) virus as a novel major variant, it too would have been classed as a pandemic.

THE DOCTRINE OF ORIGINAL ANTIGENIC SIN AND OTHER UNEXPECTED FINDINGS

In the 1950s the physician attempting to make a firm diagnosis of influenza had either to isolate the virus or to demonstrate an increase in the circulating antibody to it in the serum of the blood of the patient. Isolating the influenza virus was a complicated procedure demanding meticulous preparation of water supply, glassware, transport medium, and other necessities. The virus needed either a culture of living cells or an intact animal in which to replicate and was eclectic as to the type of host cell chosen for its cultivation. The appropriate cells were difficult to prepare and maintain and were often already harboring bacteria, fungi, or other viruses. The use of chicken embryos inside fertilized eggs or of laboratory animals avoided some of the difficulties of cell culture but presented their own problems and they were expensive. Not surprisingly, most influenza diagnoses were serological, demonstrating that a rise in circulating antibody had occurred between an early and a later specimen of blood from the patient.

At first no one doubted that a rise in antibody against, for example, A(H0N1) influenza virus or a single specimen showing a high level of such antibody proved that the recent illness had been influenza caused by an A(H0N1) strain of the virus. Even now such a finding in a symptomless patient is usually accepted as evidence that he has had a subclinical infection with the virus against which the antibody is directed.

In 1953, Davenport, Hennessy, and Francis,[2] working in the United States, showed that such simple interpretation of influenza serology could be misleading. In children the range of antibody spectrum is narrow but it becomes broader in later life, and the authors found a correlation between the periods during which certain strains of influenza virus had been prevalent and the age of persons in whom the strain-specific antibodies were currently found. Therefore:

> The antibody-forming mechanisms appear to be orientated by the initial infections of childhood so that exposures in later life to antigenically related strains result in a progressive reinforcement of the primary antibody. The highest cumulative antibody levels detectable in a particular age group tend, therefore, to reflect the dominant antigens of the virus responsible for childhood infections of the group. Hence the pattern of antibody distribution determined currently in age groups provides a serologic recapitulation of past infection with antigenic variants of the influenza viruses.

In 1955, Davenport *et al.*[3] added:

> . . . not only is the antibody mechanism oriented by the initial infection of childhood, but . . . with experience, antibody reacting with successively prevalent strains is added. This results in a continuing broadening of the antibody spectrum with age which confers the immunity of the older age groups in the population.

(See also Serious Difficulties in Explaining Antigenic Drift, Chapter 9.) They then made a remarkable observation concerning some anachronistic findings that are pertinent to discussions about possible explanations of the nature of antigenic shift in Chapter 10:

> The analysis of the antibody levels of individual sera showed a few individuals in childhood and adolescence also to have antibody against major antigens of viruses which were no longer prevalent. This shows that strains with major antigens of older viruses have had a limited circulation in recent years.

Although that seems to be the reasonable explanation, it too requires explanation of how nonprevalent strains could continue a limited circulation. At this point we should mention the startling observations of Henle and Lief,[4] also working in the United States, published in 1963:

> Consecutive infections of guinea-pigs, mice and man with one or several strains of influenza type A virus gradually lead to the appearance of antibodies to homotypic strains to which the animals were not—and the humans could not have been—exposed. The broad spectra of antibodies detected by complement-fixation were also demonstrable by neutralization and protection tests. It is tempting to speculate that multiple exposures of man to live attenuated virus vaccines at appropriate intervals may likewise lead to a broad antibody spectrum which may include not only antibodies to strains of the past but of the future as well.

They explained their findings as follows:

1. All virus *populations* may contain a few aberrant virus particles with antigenic patterns of earlier or future stains.
2. All virus *particles* may contain antigenic determinants of most or all strains of a given type. (The dominant particles or determinants identify the strains of given years. The trace particles or determinants are detectable only by the fact that they recall antibodies, gradually evoke antibodies after several exposures to them; or induce antibody formation after overwhelming infections in which they may attain an effective mass.)
3. On the other hand, in response to a severe primary infection and especially to repeated attacks of influenza, some antibodies may be formed which are capable of reaction with a number of V antigens.

They then provide a list of the possible responses to different sorts of challenge:

1. In primary infections: antibodies to infecting virus only.
2. Overwhelming primary infections: broad antibody spectrum.
3. Subsequent infections: homotypic antibodies to previous infections.
4. Multiple infections: antibodies also to strains never encountered.

Thomas Francis, Jr, one of the discoverers of influenza B virus, gave a paper in 1960 to the American Philosophical Society brilliantly entitled "On the Doctrine of Original Antigenic Sin."[5] He summarized the doctrine as follows:

> The antibody of childhood is largely a response to the dominant antigen of the virus causing the first Type A influenza infection of the lifetime. As the group grows older and subsequent infections take place, antibodies to additional families of virus are acquired. But the striking feature is that the antibody which is first established continues to characterize that cohort of the population throughout its life. The antibody-forming mechanisms have been highly conditioned by the first stimulus, so that later infections with strains of the same type successively enhance the original antibody to maintain it at the highest level at all times in that age group. The imprint established by the original virus infection governs the antibody response thereafter. This we have called the doctrine of original antigenic sin.
>
> The effect is attributed not merely to continuation of initial antibody levels but to repeated stimulation by *persistence of the first dominant antigen* as a lesser or secondary component of later Type A strains.

Some later workers reported that original antigenic sin operates only between certain of the type A influenza virus subtypes. It has been said that H2 and H3 antigens do not evoke H1 antibody in persons and animals primarily infected with A(H1N1) strains, though not all workers agree.

SEROLOGICAL ARCHAEOLOGY AND THE DISCOVERY OF RECYCLING OF ERAS OF PREVALENCE

Soon after the great antigenic shift of A(H1N1) to A(H2N2) in 1957, two workers in the Netherlands, Professor J. Mulder and Dr. Nick Masurel,[6] examined hundreds of sera obtained before the epidemic from persons in various parts of the country. Antibody against the novel virus (which had not arrived at the time the specimens were collected) was found to be already present in some sera. The greatest number of persons already possessing antibody against this "Asian influenza" and those with the highest content of it were persons aged 71 years or older. The authors therefore suggested that the 1957 influenza A virus must have had a previous epidemic prevalence in the last quarter of the nineteenth century, possibly that of 1889–90, which was also alleged to have first appeared in Asia. The authors predicted that similar investigations of specimens taken before a future antigenic shift of this Asian 'flu virus might substantiate their theory, and indeed their prophecy was fulfilled when the Asian A(H2N2) virus was superseded in 1968 by the "Hong Kong" A(H3N2) strains. In 1973, Masurel[7] working with W.M. Marine, produced serological evidence that "viruses with Asian/57/like and Hong Kong/68-like hemagglutinins occurred in the same sequence at the end of the 19th century as was seen in 1957 and 1968." They added the prediction that a swinelike influenza A virus might recur in mankind by 1985 to 1990, a grim prophecy because this is thought to be the pandemic virus of 1918. The latter prediction seemed about to be fulfilled when in 1976 influenza that broke out

TABLE 6.1. The Recycling of Eras of Prevalence of Major Variants of Influenza A Virus

Era	Major serotype
1889–1900	A(H2N2)
1900–18	A(H3N2)
1908–18	A(H1N1 old style)
1918–28	A(H1 swine-like)
1929–46	A(H0N1)
1946–57	A(H1N1 old style)
1957–68	A(H2N2)
1968–	A(H3N2)
1976 (no era)	A(H1N1 swine-like)
1977–	A(H1N1 old style)

among recruits at Fort Dix, Iowa, was found to be caused by the swinelike influenza A virus.[8] Fortunately it did not spread (see The Fort Dix Influenza Epidemic in Chapter 13).

The very next year, however, the serological evidence for the phenomenon of recycling eras of prevalence of major variants of human influenza A virus received virological confirmation. In 1977, the epidemic of "Russian flu" that swept through the world population was found to be caused by the Scandinavian strain of A(H1N1 old style) influenza virus that had caused a worldwide epidemic in 1951 and had not been prevalent since 1953.[9] Masurel and his colleague R.A. Heijtink[10] were also able to show serologically that the A(H1N1 old style) strain had had a still earlier era of prevalence form 1908 until about 1918 when, as at present, it had co-circulated with A(H3N2) strains (see Table 6.1).

The phenomenon of recycling of eras of prevalence is a feature that needs to be considered in relation to concepts purporting to explain the mechanism of antigenic shift. Somehow and somewhere the genome or the whole virus particles are being stored between successive eras of prevalence.

REFERENCES

1. Langmuir AD, Schoenbaum SC: The epidemiology of influenza. *Hosp Pract* 11:49–56, 1976.
2. Davenport FM, Hennessy AV, Francis T: Epidemiologic and immunologic significance of age distribution of antibody to antigenic variants of influenza virus. *J Exp Med* 98:641–656, 1953.
3. Hennessy AV, Davenport FN, Francis T: Studies of antibodies to strains of influenza virus in persons of different ages in sera collected in a postepidemic period. *J Immunol* 75:401–409, 1955.
4. Henle W, Lief FS: Broadening of antibody spectra of antibodies to influenza. *Am Rev Respir Dis* 88(3):379–386, 1963.

5. Francis T: On the doctrine of original antigenic sin. *Proc Am Philosoph Soc* 104:572–578, 1960.
6. Mulder J, Masurel N: Pre-epidemic antibody against 1957 strain of Asiatic influenza. *Lancet* 1:810–814, 1958.
7. Masurel N, Marine WM: Recycling of Asian and Hong Kong influenza A virus hemagglutinins in man. *Am J Epidemiol* 97:44–49, 1973.
8. Masurel N: Swine influenza virus and the recycling of influenza A virus in man. *Lancet* 2:244–247, 1976.
9. Kendal AP, Noble GR, Skehel JJ *et al*: Antigenic similarity of influenza A(H1N1) viruses from epidemics in 1977–1978 to "Scandinavian" strains isolated in epidemics of 1950–1951. *Virology* 89:632–636, 1978.
10. Masurel N, Heijtink RA: Recycling of H1N1 influenza A virus in man. *J Hyg* (Camb.). 90:397–402, 1983.

7

The Necessity for a New Concept

INTRODUCTION

In previous chapters we described how the behavior of epidemic influenza puzzled our medical forebears and how the sudden explosion of information about the nature of the virus and its antigenic reactions in the human host have added to the features calling for explanation. In this and subsequent chapters we shall examine in greater detail other aspects of its behavior that defy explanation by the current concept of direct spread and discuss modifications that have been introduced and alternative hypothesis. Some alternatives, while answering particular difficulties, run afoul of others. A theory of epidemic influenza needs to be unifying, explaining all the epidemiological problems, and the new concept will be examined and tested for such potentiality.

INTEREPIDEMIC ABSENCE OF THE PREVALENT INFLUENZA VIRUS

A typical influenza epidemic is unmistakable and the virus is usually easily isolated from the patients at such times. The virus, however, virtually disappears in the long interval between one epidemic and the next. The problem of its absence became apparent soon after the discovery of human influenza virus, and in 1945 Burnet[1] was compelled to conclude that the virus must be surviving in some mode of latency within the tissues of human carriers. He had been investigating human *Herpes simplex* virus, which causes a severe general infection when it first invades children and leaves a lifelong residue of latent virus that is periodically reactivated, usually as "cold sores" from which nonimmune companions can catch their primary herpetic illness. Burnet speculated that influenza virus might have a similar though not identical natural history. Andrewes[2] too conceded reluctantly that the interepidemic absence of influenza virus must betoken some mode of latency.

As time passed microbiologists became more familiar with the behavior of the many other families of viruses, and it seemed probable that latency was not a feature in the natural history of the Orthomyxoviruses, although it was characteristic of most herpes viruses. Even today many microbiologists doubt if influenza viruses exist in any mode other than the acute replicative infection similar to that of measles virus, a view incompatible with the new concept. The question of persistent infection by influenza virus will therefore be discussed later.

The current concept states that influenza virus persists between epidemics by continuous person-to-person spread, albeit at a low level involving numerous asymptomatic infections. Influenza is clinically indistinguishable from numerous other respiratory virus infections and tends to be underdiagnosed in the absence of an epidemic. In 1974, Dowdle[3] reported at an influenza workshop that he had been able to isolate the virus from influenzal infections in every calendar month, and ten years earlier Dingle and his colleagues[4] in Cleveland, Ohio had found that some healthy persons had shown antibody increases in their sera during the interepidemic period of a surveillance.

These and comparable findings impressed Kilbourne,[5] who pointed out that people with symptomless infections would be mixing normally in the community and able to transmit the virus widely. He claims that these silent infections form an invisible portion of the endless chain of direct transmissions that links successive epidemics and so secures survival of the influenza virus. He makes the further suggestion that the number of persons with symptomless influenza increases *pari passu* with the increasing proportion of immune persons in the community, though he provides no evidence and attempts no explanation. In 1987, he wrote: "Thus, the survival of virus between epidemics can be explained as the result of its continuing sequential transmission from person to person at a level below the epidemic threshold."

Kilbourne's statement is a succinct account of the current concept, but it is inadequate in a number of ways. He himself admits that there is no evidence that virus can be transmitted by persons with symptomless influenza, and the interepidemic cases of the disease are so scanty that they could hardly provide the required endless chains of transmission. Were they more numerous during the long interepidemic months, insufficient nonimmune persons would escape to provide the subsequent epidemic. The most serious omission from this hypothesis is the complete absence of any attempt to explain why influenza should suddenly switch from high epidemicity to inapparent epidemicity and then, months later, revert to epidemicity, repeating the process year after year.

The new concept proposes that, as Burnet and others have suggested, the virus remains noninfectious in some mode of persistence or latency in persons recovered from influenza. Such symptomless carriers would be widely disseminated throughout the community and would only become infectious when a seasonally mediated stimulus reactivated the virus in their tissues. The apparent interepidemic absence of the virus would thus be simply explained.

Von Magnus[6] first described incomplete virus particles that are produced early in influenzal infection and have the property of interfering with the replication of standard infectious virions. They are used in the laboratory to induce persistent noncytopathic infection of cell cultures, and they are now known as *defective interfering particles* (DIPs) because even in very small numbers their presence can completely prevent the production of standard infectious virus. Because DIPs are produced early during natural infections, the new concept suggests that they may form a part of the mechanism that switches the virus from epidemicity to nonepidemicity by producing the interepidemic carrier state that explains the survival of the virus during its apparent absence between epidemics. The persistence of influenza virus and DIPs has frequently been demonstrated in cell cultures.

Latent or persistent virus has not yet been found in human carriers, but is being sought by modern techniques of molecular virology. The new concept offers a credible, if tentative, alternative to the current concept.

WHY DO INFLUENZA EPIDEMICS CEASE?

The next problem to be considered is connected with the previous problem, namely, why do epidemics terminate in situations that are admirably suited for their continued spread? For example, when in 1957 A(H2N2) strains everywhere replaced the A(H1N1) strains that had been causing all the influenza A for many years, both the hemagglutinin and the neuraminidase had been changed so that few people possessed immunity against the novel viruses. The world population was therefore wide open to attack by this "Asian 'flu," except (as was later discovered) persons over 70 years old some of whom retained immunity from attacks sustained during a previous era of prevalence of a similar subtype in the last quarter of the nineteenth century. It was correctly assumed that most persons in the British Isles would possess little protection against this new virus.

The general practitioners had been enjoying an exceptionally quiet winter in 1956–57 and had experienced little influenza among their patients. In February 1957, news reached Britain that a novel influenza A virus was causing thousands of cases of influenza in southeast Asia against which neither a previous attack of influenza A nor the current vaccine were providing any protection. The progress of the epidemic across the globe was watched with apprehension. Nobody doubted that it would reach this country, an expectation that itself provides food for thought.

Several months elapsed before this "Asian" A(H2N2) virus was found in the

British Isles. In late May and early June 1957, the virus was isolated from cases of influenza during an outbreak in a military establishment in Cheshire. The novel virus had secured a foothold in the nonimmune population of Britain and must surely, one supposed, proceed to cause a nationwide epidemic. However, the unseasonable outbreak ceased and no further cases were recorded in Great Britain throughout that summer.

What stopped it? One cannot imagine measles failing to take advantage of such a situation, and here was another exceedingly infectious agent that had gained a good foothold in an almost immunologically virgin community and yet it failed to spread more widely. The phenomenon needs to be explained and we shall later look more closely at such out-of-season epidemics (Chapter 17).

In the community of Cirencester, Gloucestershire in 1957, few acute respiratory illnesses demanded attention until September. In late August, influenza caused by A(H2N2) strains was reported from Northern Scotland, but despite close surveillance there was no sign of the presence of the disease in Cirencester until 23 September. Then, as shown in Figure 7.1, Asian 'flu erupted in the community. In the space of three weeks it had attacked some 8% of our population and was therefore in the ideal situation to attack the remainder of the largely nonimmune community. The influenza patients were widely distributed so that almost everybody was at risk of contracting the infection if it was being spread from the sick. Measles, similarly placed, could not fail to attack almost every susceptible person

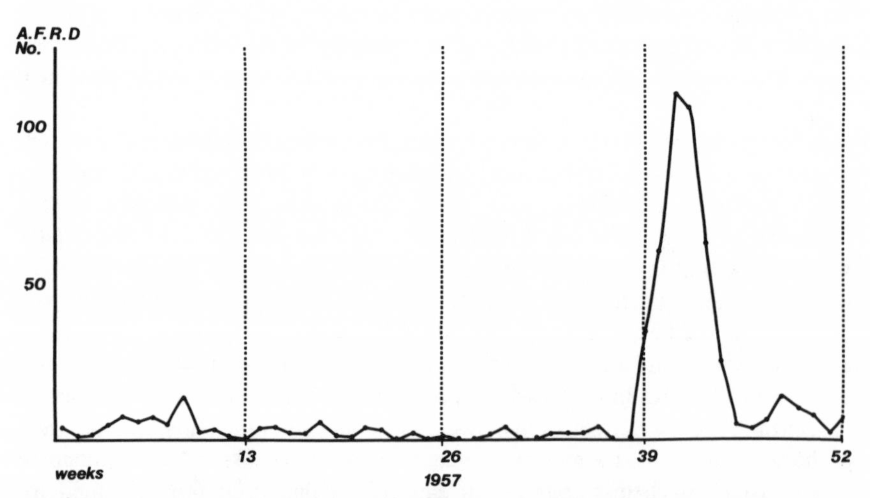

FIGURE 7.1. Acute febrile respiratory diseases (AFRD) treated in the Cirencester general practice in 1957 to show the explosive impact of the first epidemic of Asian A(H2N2) influenza virus at the end of September and October. What stopped it after only six weeks in a largely nonimmune community? (From Hope-Simpson,[14] p. 34, Fig. 1; reproduced with permission from *PHLS Microbiology Digest.*)

as was demonstrated when it invaded virgin communities in the Solomon Islands and in the Faroës.[7]

Asian influenza virus failed to take advantage of the opportunity. The third week had witnessed the peak of the epidemic, and it waned as rapidly as it had arisen and was over after six horrible weeks from its beginning, having attacked about 15% of our community. How did it so rapidly decline and cease in conditions ideal for it to continue by direct transmissions from the numerous and ubiquitous sufferers distributed widely among the remaining nonimmune persons who probably still comprised more than 70% of the community? No explanation of the familiar phenomenon seems to have been offered by those who favor the hypothesis of direct spread, and indeed an explanation along those lines is difficult to find.

The new concept provides a simple explanation by suggesting that the sick cannot usually transmit the virus during their illness. The epidemic therefore consists entirely of persons who have caught the virus from carriers (infected in a previous epidemic) in whom persistent noninfectious virus has been briefly reactivated. If the virus cannot be transmitted from the influenzal patient, every epidemic must cease automatically as soon as such patients have had their attack of influenza, irrespective of the number and availability of their nonimmune companions. Examples such as this from epidemics occurring soon after an antigenic shift are valuable because of the high proportion of nonimmune persons in the community.

Another example is provided by the subsequent antigenic shift. The era of unchallenged prevalence of A(H2N2) strains came to an end when A(H3N2) strains appeared in Hong Kong in July 1968. "Hong Kong 'flu" immediately replaced Asian 'flu and in Great Britain we once again awaited with trepidation the arrival of the novel subtype, not doubting that it would come.

No influenza occurred in the Cirencester community through the summer and autumn of 1968. In mid-December a solitary case occurred in a 15-year-old boy sent home from a boarding school 50 miles away to avoid Hong Kong influenza that had broken out in the school. He had mixed in the Cirencester community for several days before he developed the disease. The A(H3N2) virus was isolated from his specimen taken on 12 December 1968. No spread occurred from this case. The epidemic in this community did not begin until more than a month later on 15 January 1969, and it lasted for 13 weeks until mid-April attacking less than 5% of our community, almost unnoticed among the usual winter ailments.

Why was this first epidemic of the era of prevalence of the new subtype so small, and what caused it to stop? It was a trivial consequence for the impact of a novel virus in immunologically virgin territory (Fig. 7.2).

Figure 7.3 and Table 7.1 show that the community was less immunologically chaste than we had supposed, because only the hemagglutinin had been changed. The old N2 neuraminidase antigen had been conserved almost unchanged from the

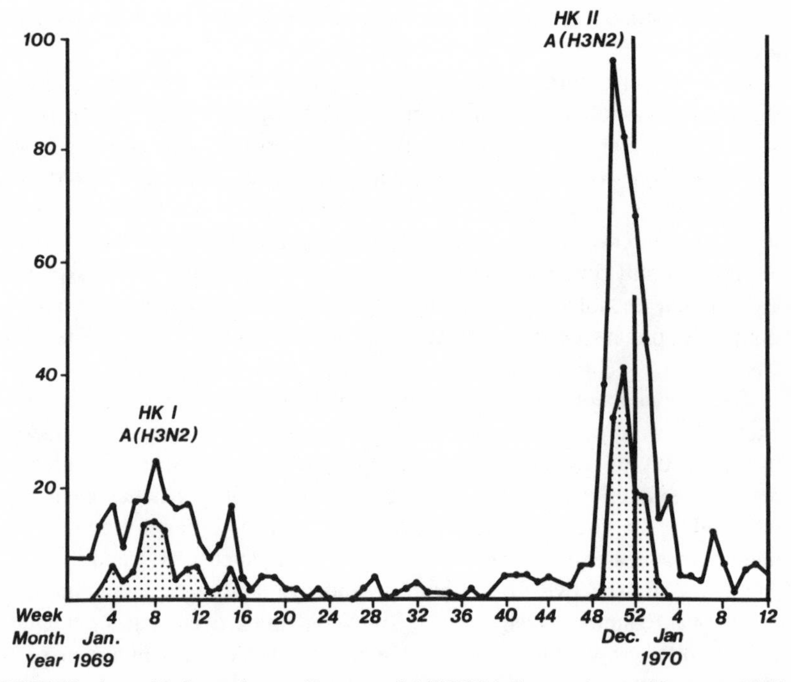

FIGURE 7.2. Acute febrile respiratory diseases and A(H3N2) influenza virus positive cases (shaded) in the first two Hong Kong influenza epidemics (HKI and HKII) in Cirencester. What stopped these epidemics in such a largely nonimmune community? (From Hope-Simpson,[14] p. 35, Fig. 2; reproduced with permission from *PHLS Microbiology Digest.*)

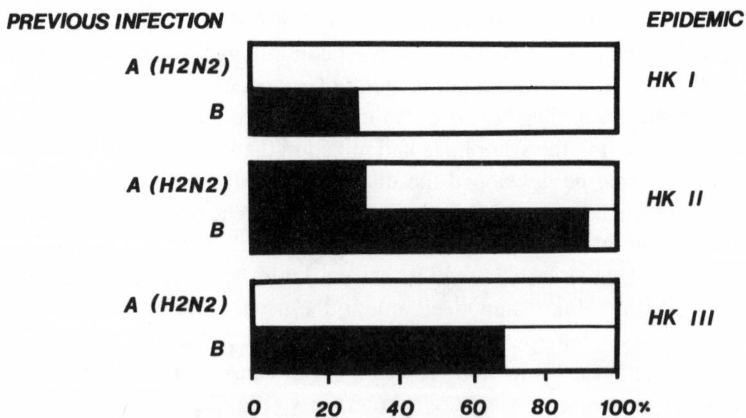

FIGURE 7.3. During the first three Hong Kong epidemics caused by A(H3N2) influenza virus, a previous attack of Asian A(H2N2) influenza conferred more protection than a previous attack of influenza B. The cases of Hong Kong influenza (shaded) are shown as a percentage of the number expected had there been no protection (based on the data in Table 7.1).

TABLE 7.1. The Number of Persons Known to Have Suffered a Previous Attack of Asian A(H2N2) Influenza Who Were Attacked by A(H3N2) Virus in the First Three Hong Kong Influenza Epidemics Compared with the Numbers of Those Previously Attacked by Influenza B Virus Who Were Later Attacked by A(H3N2) Virus[a]

Challenged by A(H3N2) virus		Previously attacked by					
		A(H2N2) virus			Type B influenza virus		
Epidemic	Attack rate	Number	Expected	Observed	Number	Expected	Observed
HKI (1968–69)	5%	118	6	0	75	4	1
HKII (1969–70)	15%	118	18	5	74	10	9
HKIII (1971–72)	4%	113	4	0	65	3	2
Total		118	28	5	75	17	12

[a] A previous attack of A(H2N2) influenza seems to have conferred more protection than influenza B virus against A(H3N2) infection (from Hope-Simpson[8]).

previous A(H2N2) strain and had conferred some protection on persons who had been infected.

Another explanation offered of the small size of the first Hong Kong 'flu epidemic was the alleged high proportion of symptomless infections, more than 40% of the population, it was said, having shown antibody increases against strains of the new subtype. The epidemic had resembled an iceberg of which the 5% of patients suffering from clinical influenza had been only the visible tip beneath which was the vast submerged portion of persons who had been asymptomatically infected.

This reasonable explanation must have been mistaken because only eight months later the community was attacked by the second A(H3N2) epidemic (Fig. 7.2), which was as explosive and formidable as that of A(H2N2) strains in 1957. The first case in our Cirencester community occurred on 5 December, the epidemic peaked within a fortnight, and was over in less than six weeks. In that short time it had attacked nearly three times as many persons as the first A(H3N2) epidemic had attacked in 13 weeks. We are again faced with the problem of explaining what stopped these two epidemics, known as HKI and HKII. There must have been many nonimmune persons unattacked after HKI and their number, though reduced, was still disproportionately high after HKII as was later shown by the subsequent epidemics caused by A(H3N2) strains during the next decade.

The new concept again seems to provide a reasonable explanation. The size of each epidemic is seen as depending entirely on the number and distribution of reactivating carriers in relation to the number and distribution of nonimmune persons. Epidemic influenza is thus seen as a seasonal crop, and the absence of direct spread from the sick explains why epidemics cease regardless of the presence of susceptible companions.

THE DIFFERING CHARACTERS OF THE EPIDEMICS OF 1957, 1968–69, AND 1969–70

The differing character of influenza epidemics is well illustrated by the first two epidemics of Hong Kong 'flu, HKI so small and protracted that the public was not aware of its presence, HKII large and explosive. The illness probably differed little in severity, but the social disruption caused by the secular intensity of HKII gave the impression of much greater severity. When many people are ill contemporaneously, there may be no one to call on for assistance and the simplest services are disrupted.

Both the 1957 A(H2N2) epidemic and the 1969–70 A(H3N2) epidemic were explosive. They erupted suddenly at a time when no influenza and little other febrile illness had been present in the community for many months. The earliest cases were simultaneous over a wide area and no communication between them could be traced. Table 7.2 gives some indication of the explosive nature of the 1957 epidemic in an area of about 10,000 square miles.

What is the explanation of differences between influenza epidemics, even those caused by viruses of the same subtype? There is no simple answer. A small mutation in the gene coding for hemagglutinin caused a major change in severity of an avian influenza virus. In April 1983, A(H5N2) influenza virus was found to be causing a mild respiratory illness in the flocks in Pennsylvanian chicken farms. Suddenly, in the following October, the virus changed its behavior and became lethal causing a generalized viremia with a high mortality among the chickens. The change had been caused by a tiny mutation. Analysis of the nucleotide sequences of the genome by R.G. Webster and his colleagues[9] found that the relatively harmless April virus had undergone a point mutation in the gene coding for the hemagglutinin causing the loss of a carbohydrate at a single site. This tiny structural change had transformed the negligible pathogen into a killer. Influenza A virus subtypes containing H5 or H7 antigen are usually strongly pathogenic, and A(H5N2) virus commonly causes severe illness in birds. It therefore seems prob-

TABLE 7.2. Asian 'Flu in Various Localities[a]

Location	Distance and direction	Onset	Peak
Rugby School	60 miles N	24.9.57	30.9.57
Marlborough School	25 miles S	24.9.57	30.9.57
Nottingham School	100 miles NE	23.9.57	30.9.57
Cirencester families	0	23.9.57	2.10.57
Peaslake School	100 miles E	24.10.57	29.10.57

[a]It erupted simultaneously over a wide area, yet had a different timing in a neighboring locality.

able that the milder April 1983 strain of A(H5N2) influenza virus was the aberrant variant.

The well-researched and well-documented experience in the chicken farms of Pennsylvania in 1983 carries a frightening lesson about the dangerous potentialities of even the mildest influenza viruses, but the difference between HKI and HKII epidemics concerned the distribution of the disease, not the pathogenicity of the virus, and could be explained by the relative distribution in the community of reactivating carriers and nonimmune persons if the new concept is correct.

The host cell exerts great influence on the influenzal parasite so that it is wise to be cautious in transferring knowledge obtained from influenzal infection of one host species to that of another. No such mutation as occurred in the avian A(H5N2) virus has been found to account for the differences of HKII from HKI in mankind. The new concept suggests that in the 1968–69 season the reactivating stimulus must have been protracted and operating on a world population thinly, though ubiquitously, seeded with carriers, and that many of the abundant "nonimmune persons," those who had never had Hong Kong influenza, carried a partial immunity from N2 neuraminidase antibody. Thus HKI was small and protracted. In the following season of 1969–70, on the other hand, the proportion of carriers would have increased by 5% of the population and the reactivating stimulus was probably short, resulting in the intensity of HKII. In some parts of the United States, the first Hong Kong 'flu epidemic was the larger of the two.

When influenza is considered as an annual crop seeded in humanity, it is not difficult to understand why some epidemics should begin explosively and contemporaneously over large areas, nor why some areas should be out of step (like the Peaslake school in Table 7.2), according to vagaries in the behavior of the seasonally mediated stimulus. Season is clearly of great importance in the behavior of epidemic influenza, and the discussion on the nature of seasonal phenomena (Chapter 8) shows the stimulus at work on a global scale. Seasonal effects on influenza differ from year to year and between one location and another. In this, influenza resembles other seasonal crops.

PROBLEMS FROM HOUSEHOLD STUDIES: LOW ATTACK RATE AND ABSENT SERIAL INTERVAL

Early news of the antigenic shift in 1968 gave time to prepare for careful household studies in the Cirencester community when the epidemic should reach the British Isles. These studies were added to the continuous laboratory and clinical surveillance that had begun in 1961 and that ended in 1976.

The prospective survey of households infected by A(H3N2) influenza virus in the first HK epidemic of 1968–69[10] found unexpectedly that the virus had failed to spread in about 70% of the households into which it had gained entry by causing a case of influenza (Fig. 7.4). Moreover, no serial interval between causally related cases could be demonstrated by cumulating household outbreaks (Fig. 7.5A).

The serial interval is epidemiologically valuable in several ways. It is a measure of the interval between a case of an infectious disease and the case that caused it. The demonstration of its presence is therefore the epidemiological evidence that the agent is being transmitted directly from the sick person to his companions in a particular environment, for example, the household. Absence of the serial interval is evidence that the disease was not being directly transmitted from the sick.

The absence of the serial interval is noteworthy. Figure 7.4 shows that where more than one case occurred in a household, subsequent cases all followed within a few days. Each such household must have been infected from a single introducing source, but the absence of a demonstrable serial interval in Figure 7.6C means that the source in each household was an asymptomatic carrier in whom the persistent virus had become briefly reactivated to infectiousness.

The suggestion of Magrassi (see Chapter 4, and reference 9) and others that the anomalous behavior of epidemic influenza may be explained by the virus lying latent in a person before his attack of influenza does not fit the findings in Figures 7.4 and 7.5A. On Magressi's hypothesis, the cases in households with more than one case should have appeared randomly throughout the long course of the HK epidemic, not closely linked as they were in each such household (see Chapter 4: Field Studies that Stimulate New Ideas).

Determination of the serial interval is also needed to identify the secondary cases, separating them from co-primaries, tertiaries, and others.[11] Without knowing the serial interval, one cannot accurately determine the secondary attack rate. Figure 7.6, A,B shows the serial interval in a school measles epidemic and in

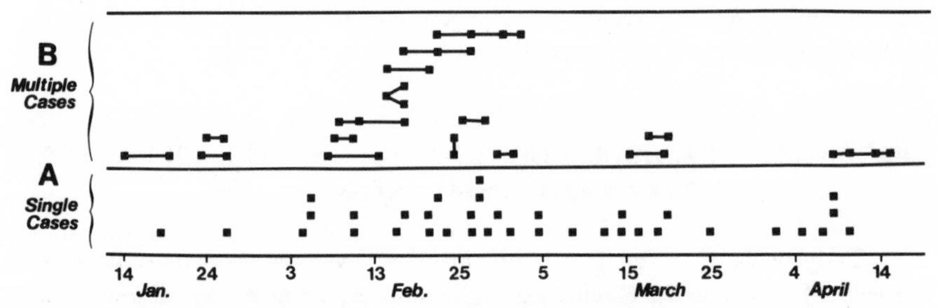

FIGURE 7.4. Household outbreaks during the first Hong Kong epidemic in Cirencester. (A) Nearly 70% of infected households had only one case. (B) In the multiply infected households, the cases were so closely linked that the virus had probably been acquired from a source within the household. Absence of a serial interval in Fig. 7.5A excludes an introducing case and suggests introduction by a symptomless carrier. Symptomless persistence cannot have preceded the illness, otherwise the multiple cases would have been randomly distributed throughout the epidemic irrespective of household.

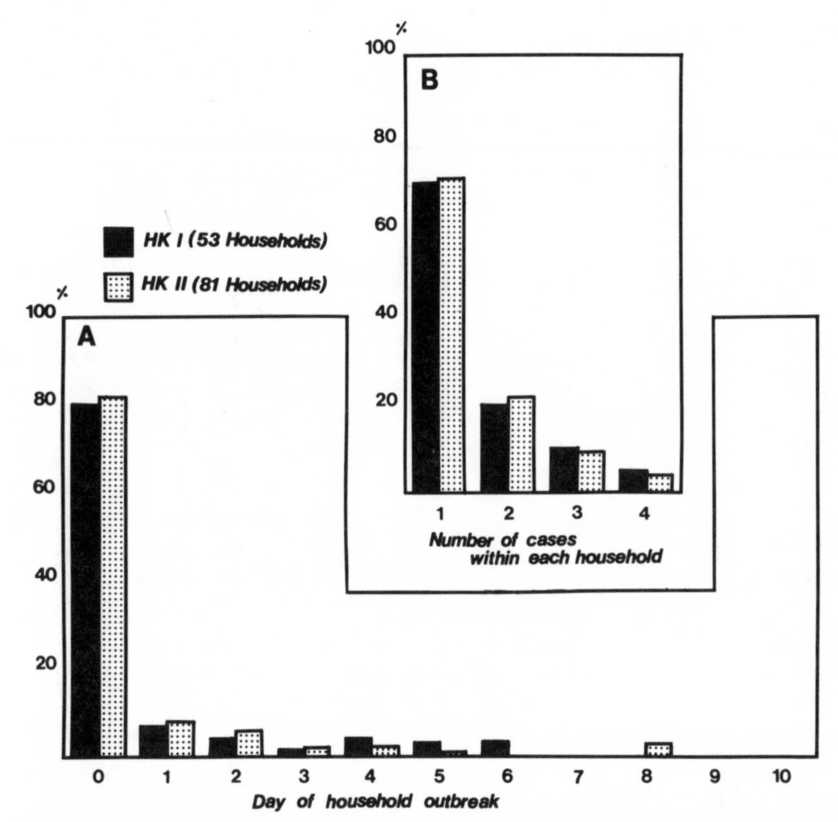

FIGURE 7.5. HK I and HK II behaved similarly in Cirencester households. (A) Percentage of cases by day of household outbreak. Note the absence of a serial interval in both epidemics. (B) Percentage of households by the number of cases in them (from Hope-Simpson,[10] p. 20, Fig. 8; reproduced with permission from *Epidemiology and Infection*).

cumulated measles households. Its absence in the Hong Kong influenza epidemics in Cirencester made it impossible to compute the true secondary attack rate of influenza virus in the infected households, an important measurement because it gives information about the infectiousness of the infecting agent. An approximation known as the "subsequent attack rate" was therefore employed, adopting any first-day case in each household as the "introducer" and all the later cases in each household as "secondaries." The method inflates the rate because it automatically includes all co-primary, tertiary, and later cases among the secondaries. Despite this inflation, the intrahousehold subsequent attack rate in HKI was found to be only 17%, a ridiculously low figure. The secondary attack rate in measles households has been found to be 75%, in varicella households 61%, and in mumps households 30–35%.[12]

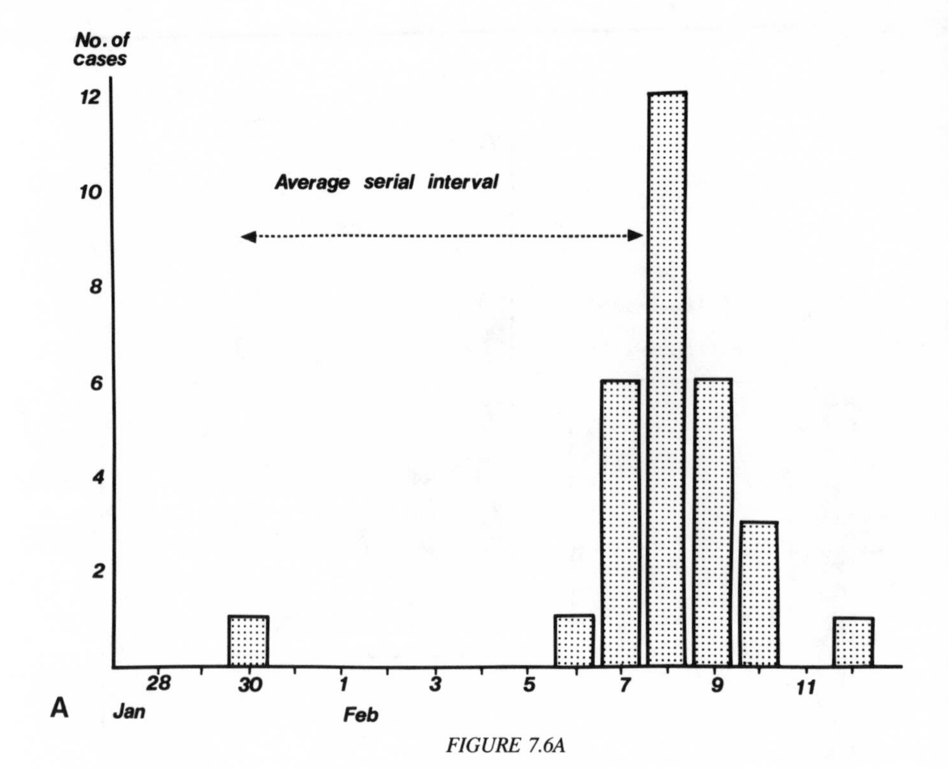

FIGURE 7.6A

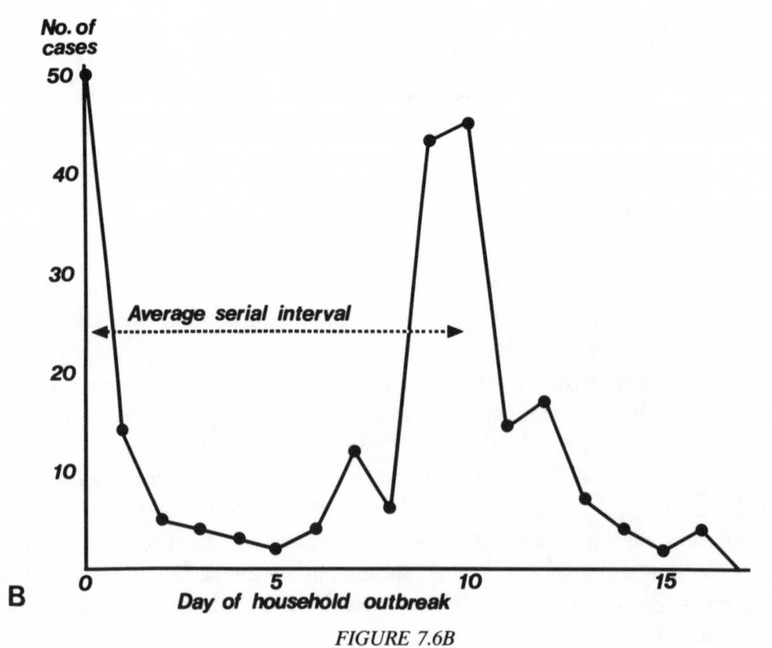

FIGURE 7.6B

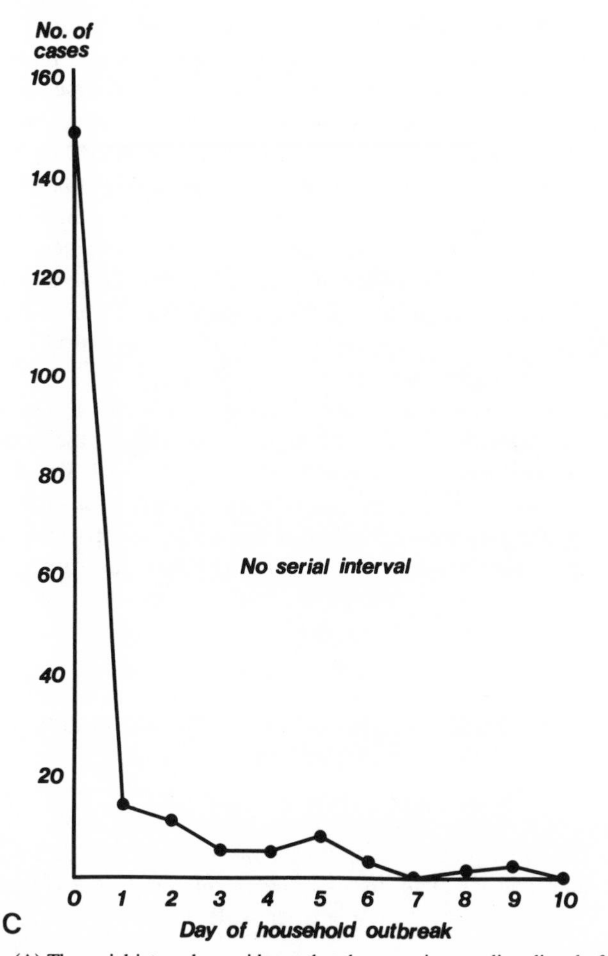

FIGURE 7.6. (A) The serial interval as evidence that the agent is spreading directly from the sick in a primary school environment (from O'Brien and Hill[15]; reproduced with permission from *PHLS Communicable Disease Report*). (B) The serial interval as evidence that measles is spreading directly in 264 cumulated households. (C) For contrast with A and B, there is no evidence of a serial interval in 134 cumulated influenza household outbreaks.

HKI had been a small, prolonged epidemic and might therefore have been in some way atypical. HKII, on the other hand, was a textbook epidemic of severe influenza, and indeed many more households were attacked than in HKI. Nevertheless, the analysis of the intrahousehold behavior in HKII might have been a duplicate of that in HKI as is shown in Figure 7.5. Here again more than 70% of affected households had only a single case, no serial interval could be demonstrated in the cumulated household outbreaks (Fig 7.5A), and the household "subsequent attack rate" was even lower, only 14%. These findings were incompatible with the current concept and the absence of a serial interval made it imperative to seek an alternative epidemiological hypothesis (see Fig. 7.6C).

The new concept proposes that the introducer of influenza virus into each household is not a case of influenza but a symptomless carrier in whom the persistent virus has been reactivated to infectiousness. In that situation the findings are readily comprehensible. None of the household cases are primaries and almost all of them are secondary to the introduction of reactivated virus in the unidentifiable carriers. No serial interval can be demonstrated because there are no primary cases. If, as thus postulated, all the household cases were secondary to the symptomless carriers, the true secondary attack rate in HKI must have been 25%, not 17%, and in HKII 55%, not 14%.

Another finding that excites frequent comment has been the contrast between the low household attack rate and the high rate caused by the same influenza epidemic in institutions such as schools and military barracks. The explanation is probably to be found in the fact that no account of primary or introducing cases is taken in computing the institutional rates, so they are comparable to the household rates as calculated by the new concept. The attack rates for HKI and HKII as determined by the new concept accord well with those caused by these epidemics in boarding schools and barracks.

EPIDEMICS IN SMALL LOCALITIES REFLECT THE NATIONAL EXPERIENCE

A feature of epidemic influenza that has puzzled many observers including Andrewes is the rapidity with which novel strains appear in different parts of the globe, a problem that we shall discuss later in relation to the global picture of influenza. Here we consider it on a smaller scale within the bounds of a single country.

If the virus is surviving by spreading directly from case to case, each novel strain must somewhere invade the country from without and become distributed in this way throughout the population. Small rural populations would therefore lag behind the central laboratory receiving specimens from all parts of the nation and would also reflect only an incomplete portion of the national experience.

The Cirencester general practice served a semirural community of about 3700 persons living in an area of less than 50 square miles. During eight successive seasons from 1968 to 1976, it closely reproduced in miniature the pattern of A(H3N2) influenza viruses received from all affected regions of England and Wales by the Central Public Health Laboratory at Colindale, London. All the strains examined were similar to the prototype strain A/Hong Kong/1/68 until 1971, but thereafter the antigenic drifts began to produce a complex pattern. Six minor antigenic variants came and went, three of them being isolated for only a single season (Table 7.3).

Season after season the small local epidemics in Cirencester reproduced the complicated changes affecting the whole country except for the failure in Cirencester to isolate one of the three variants present in 1974–75. It is difficult to understand how such concordance could be achieved by viruses invading the United Kingdom anew each season from outside its borders and traveling to all parts of the country by chains of transmission of which each link must be a patient sick with influenza. The most remarkable example of this problem occurred during the winter of 1975–76 when the novel strain A/Victoria/3/75, first isolated at Colindale on 31 December 1975, was found next day in Cirencester on New Year's day, 1976. The most logical explanation must involve some system whereby the novel strains are being produced locally, and the same reasoning must apply to the behavior of epidemic influenza in all other countries throughout the world.

The supposition of the new concept that symptomless carriers are always ubiquitously distributed throughout the world population and that a seasonally mediated stimulus reactivates these dormant viral parasites to become infectious would explain how the influenza A virus was able to appear season after season throughout the United Kingdom without recurrent annual invasions from foreign sources. It does not at first sight explain the synchronicity of the successive antigenic variations at national and local centers, but an explanation of the phenomenon is given later in Chapter 9 describing the application of the new concept to the problems of antigenic drift.

TABLE 7.3. Cirencester Isolates of Influenza A Virus Compared with the Colindale Isolates Received from All Parts of England and Wales, 1968–76[a]

Winter	Subtype A(H3N2) strains	No. of Cirencester isolates		No. of Colindale isolates	
1968–69	A/Hong Kong/1/68	3 (+73)		881	
1969–70	A/Hong Kong/1/68	(114)		800	
1970–71	A/Hong Kong/1/68	0		51 (no epidemic)	
1971–72	A/Hong Kong/1/68	54		751	
1972–73	A/England/42/72	26		1290	
1973–74	A/Port Chalmers/1/73	9		575	
1974–75	A/Port Chalmers/1/73	25	31 (+2)	241	941
	A/Intermediate/74	0		343	
	A/Scotland/74	6		347	
1975–76	A/Victoria/3/75	15	17 (+2)	1923	2041
	A/England/864/75	2		118	

[a]Number of isolates not fully characterized are bracketed (from Hope-Simpson,[10] p. 14, Table 3: reproduced with permission).

ANOMALIES IN THE AGE DISTRIBUTION OF INFLUENZA PATIENTS

In any community the ages of the persons attacked by specific infectious diseases can provide valuable epidemiological information.[13] An immunizing disease at its first introduction into a nonimmune community will attack indiscriminately persons of all ages so that the average age of the patients will approximate to that of the whole community. In successive epidemics younger persons who have been born since the first epidemic or who have reached school age and so become more accessible to the disease will comprise a higher proportion of the patients so that the average age of persons attacked will be much lower. The actual average age will be governed by a number of factors of which the most important is the degree of urbanization of the community, but in any given community the mean age of the victims will be inversely related to the infectiousness of the agent as measured by the attack rate within the household and, to less extent, to the duration of the serial interval. The higher the infectiousness and the shorter the serial interval the younger, on average, will be the patients. The relationships are well shown in Table 7.4 for the sufferers from measles, varicella, and mumps in the Cirencester community, but it is evident that type A influenza fails to conform to the rule.

After the antigenic shift in 1957 when A(H2N2) strains were replacing A(H1N1 old style) strains as the dominant influenza A virus, they attacked persons of all ages as measles would also have done in such a nonimmune community. During the following decade until 1968 there were seven subsequent epidemics caused by A(H2N2) strains in the same community. Figure 7.7 shows that the mean age of the patients from whom the virus was isolated remained high in each of the last four of these epidemics. The final A(H2N2) influenza epidemic was second only to the 1957 epidemic in size and it lasted for the first four months of 1968. Second attacks during the era of prevalence were uncommon, around 2%,

TABLE 7.4. The Relationship between the Mean Age of Persons Attacked in the Cirencester Community by Various Infectious Agents and Their Infectiousness and Serial Intervals

Infectious agent	Infectiousness	Serial interval (days)	Mean age of patients (years)
Measles virus	75%	10-11	5½
Varicella virus	61%	14	6½
Mumps virus	30%	17-18	12
Influenza A virus	?	?	35
The whole community	—	—	37

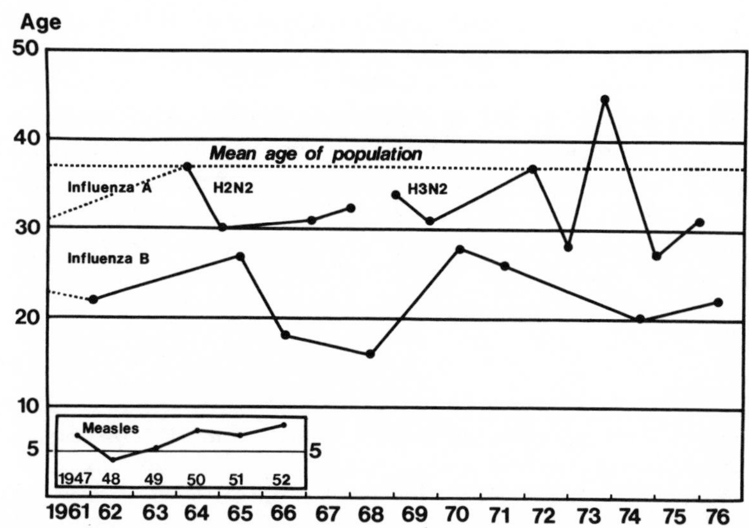

FIGURE 7.7. The average age of persons attacked in the Cirencester general practice in successive epidemics of influenza A and B compared with the age of measles patients in successive epidemics in the same area (from Hope-Simpson,[13] p. 310, Fig. 2; reproduced with permission from *Epidemiology and Infection*).

TABLE 7.5. The Average Age of Those Attacked by Infuenza A(H2N2) Virus, Influenza A(H3N2) Virus, Influenza B Virus, and Measles Virus in Sequential Epidemics in Cirencester

Influenza A(H2N2)		A(H3N3)		Influenza B virus		Measles	
Date	Average age	Date	Average age	Date	Average age	Date	Average age
1963–64	37	1968–69	34	1961/62	22	1947	6
1964–65	30	1969–70	31	1965	27	1948	4
1966–67	31	1971–72	37	1966	18	1949	5
1967–68	34	1972–73	28	1968	16	1950	7
		1973–74	45	1970	29	1951	6
		1974–75	27	1971	27	1952	8
		1975–76	31	1973–74	20		
				1976	21		
33.0 years		33.3 years		22.5 years		6.0 years	

yet the average age of the patients in that last epidemic was close to that of the community as a whole, the youngest patient being a ten-month-old girl and the oldest a man aged 86 years (see Table 7.5).

An ideal model of part of an influenza A virus era of prevalence was constructed according to the epidemiology proposed by the new concept. It showed, if the model was correct, that the average age of patients in successive epidemics would not be expected to depart widely from that found by a random sampling of the nonimmune portion of the community (reference 13, pp. 314, 315, Table 5).

REFERENCES

1. Burnet FM: *Virus as Organism* (The Edward K Dunham Lectures for 1944). Cambridge, Mass, Harvard University Press, 1945, p 105.
2. Andrewes CH: The epidemiology of influenza in the light of the 1951 outbreak. *Proc R Soc Med* 44:803–804, 1951.
3. Dowdle WR: Discussion in Epidemiology of influenza: Summary of influenza workshop IV, Fox JP, Kilbourne ED (eds). *J Infect Dis* 128:361–386, 1973.
4. Dingle JH, Badger GF, Jordan WS J: *Illness in the Home. A Study of 25,000 Illnesses in a Group of Cleveland Families*. Cleveland, Western Reserve University Press, 1964, pp 142–187.
5. Kilbourne ED: *Influenza*. New York, Plenum Medical, 1987, p 277.
6. von Magnus P: Propagation of the PR8 strain of influenza virus in chick embryos. II. The formation of "incomplete" virus following inoculation of large doses of seed virus. *Acta Pathol Microbiol Scand* 28:278–293, 1951.
7. Panum PL: *Observations Made during the Epidemic of Measles on the Faroe Islands in the year 1846*. New York, Delta Omega Society, 1940.
8. Hope-Simpson RE: Protection against Hong Kong influenza. *Br Med J* 4:490, 1972.
9. Webster RG, Kawaoka Y, Bean WJ: Molecular changes in A/chicken/Pennsylvania/83 (H5N2) influenza virus associated with acquisition of virulence. *Virology* 149:165–173, 1986.
10. Hope-Simpson RE: Epidemic mechanisms of type A influenza. *J Hyg* (Camb.) 83:11–26, 1979.
11. Hope-Simpson RE: The period of transmission in certain epidemic diseases: An observational method for its discovery. *Lancet* 2:755–769, 1948.
12. Hope-Simpson RE: Infectiousness of communicable diseases in the household (measles, chickenpox, and mumps). *Lancet* 2:549–564, 1952.
13. Hope-Simpson RE: Age and secular distribution of virus-proven influenza patients in successive epidemics 1961–1976 in Cirencester: Epidemiology and significance discussed. *J Hyg* (Camb.) 92:303–336, 1984.
14. Hope-Simpson RE: Simple lessons from research in general practice. Part 6. *PHLS Microbiol Dig* 7:34–37, 1990.
15. O'Brien J, HIIA: Outbreak of measles in a primary school. PHLS Communicable Diseases Report (CDR) 39:Fig. 1, 1988.

8

The Influence of Season

THE OMISSION OF SEASON FROM CONCEPTS OF THE EPIDEMIOLOGY OF INFLUENZA

For a long time we omitted from our conceptions the obvious and important phenomenon that influenza is a seasonal disease, possibly because no attempt seems to have been made to explain the seasonal character of influenza by the current concept of direct spread. Figure 8.1 shows the seasonal nature of epidemic influenza as experienced in a small local community in Cirencester, Gloucestershire, England, between 1946 and 1974. It was almost an annual visitant in the colder months, 22 epidemics culminating during the first quarter of the year and four in the last quarter. Only four winters saw no influenza, though a similar systematic monitoring of much larger communities would have revealed its presence every winter. No epidemics occurred during the warmer months.

Most people, especially general practitioners, are familiar with this local episodic behavior of influenza in the colder months and the Cirencester experience could be paralleled from small communities throughout the world. Figure 8.2 shows epidemic influenza appearing seasonally in Houston, Texas from 1974 to 1983.[1] Note the limited duration of each influenza season contrasting with the continuity of undifferentiated acute respiratory diseases in the lower part of the illustration.

Possibly the very familiarity of the seasonal visitations has caused the phenomenon to be taken for granted and in no need of explanation, but an influence of such regularity and importance needs to be integrated into any attempt to explain influenzal behavior. Not only influenzal epidemics but also variations in the antigenicity of the virus occur seasonally, and the problems of seasonal control are so interesting that they merit the whole of this chapter. They may well provide the key to understanding most of the influenzal problems confronting us, and the new concept assumes that it is a seasonally mediated stimulus that recalls persistent noninfectious virus to infectiousness in human carriers.

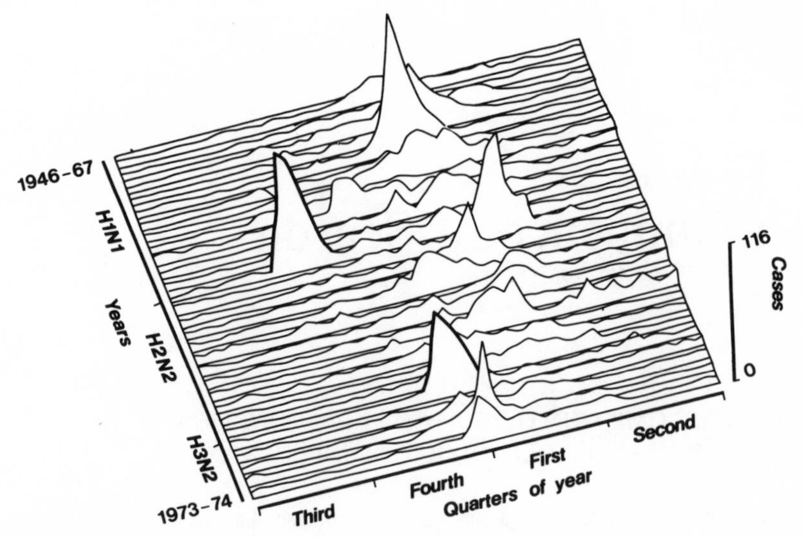

FIGURE 8.1. Acute febrile respiratory diseases treated in general practice, Cirencester, 1946–74. Of 26 influenza epidemics the peak occurred in the first quarter of the year in 22 and in the last quarter in four (from Cliff *et al.*,[9] p. 65, Fig. 3.12; reproduced with permission from Pion Limited).

THE GLOBAL VIEW OF INFLUENZA

Before considering the nature of seasonal phenomena in general, it will be valuable to have a look at the behavior of epidemic influenza on a global scale. It presents a very different picture from the parochial view of brief annual epidemics, though it makes them comprehensible.

Cirencester in England and Houston in the United States are both situated in temperate regions above the 23.5° of latitude that bounds the northern Tropics. Influenza visits the temperate regions south of the Tropics in their colder months, namely April to September. Cold, inclement weather has often been blamed for causing influenza because most of the published studies have come from workers in the areas north or south of the Tropics. Climate does not, however, determine the behavior of epidemic influenza as was astutely recognized by Sir John Pringle[2] in the eighteenth century. In a letter to Dr. John Fothergill he wrote:

> I think you do well to record the state of the weather but I think the conclusion ought to be, that the sensible qualities of the air had, most probably, no share in producing this epidemic [of 1775]. I should be tempted to say, that they evidently had no part; for we hear of the same distemper having been in Italy, France and the Low Countries, and, I doubt not, in other parts of Europe, had we inquired. But it cannot be supposed that the state of the atmosphere, either as to weight, heat, or moisture, was the same everywhere. And in the same country have we not seen it rage in one district or city, whilst others, at no great distance, were totally free. Yet between the sound and the sickly there could be no considerable meteorological difference. My

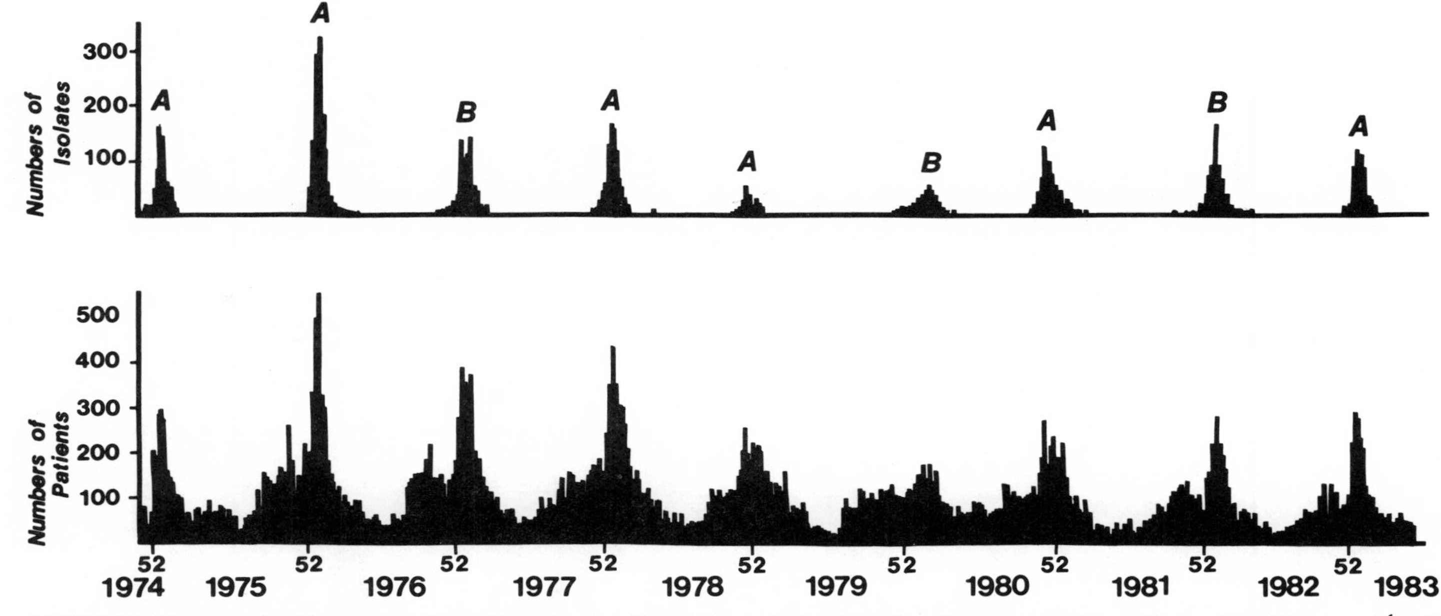

FIGURE 8.2. The seasonal epidemicity of isolates of influenza A and B and the continuity of undifferentiated acute respiratory diseases (from Glezen *et al.*,[1] p. 19, Fig. 1; reproduced with permission from Academic Press–London).

> conclusion, therefore, should be, that such epidemics (of which there have been four in my remembrance) do not depend on any principles we are yet acquainted with, but upon some others, to be investigated, and by such means as Dr. Fothergill very properly and most commendably proposes to be done by the united inquiries of this brethren.

Such letters and other contemporary writings show that influenza epidemics were then widespread over Europe and even farther afield. The slow, scanty, and dangerous human communications in the eighteenth century presented no barrier to the rapid extension of influenza epidemics.

Climate and a number of other influences may moderate or exacerbate the illness but they are not determinants of epidemic spread. Wade Hampton Frost and Mary Gover[3] in Baltimore, who believed that both common colds and influenza are spread by contagion from the sick, were compelled to discount the effect of climate. Writing 150 years after Pringle they said:

> Considering the wide geographic dispersion of the localities represented [in studies among students at widely separated universities in the USA], and their corresponding difference in climate, this uniformity of attack rate is one of the most interesting and significant facts brought out by these records, indicating that, in the prevalence of this group of disorders, climate is a factor of much less importance than would be supposed.

(See Fig. 8.3.) Influenza is ubiquitous. Epidemics occur seasonally in parts of the globe where winter as a cold season does not occur. Its global behavior can be reconstructed from the reports from many parts of the world received by the influenza center of the World Health Organization (WHO) at the Palais des Nations at Geneva. An analysis of these reports from 1964 to 1975 showed that no meaningful pattern emerged when the places of origin were classified by longitude. When, on the other hand, they were classified by the latitude of their origin, a clear pattern could be seen. Table 8.1 and Figure 8.4 represent the analyses of the reports from four major latitudinal zones of the Earth's surface, broadly as follows: (1) north of the Tropics; (2) the north Tropics; (3) the south Tropics; and (4) south of the Tropics. The table shows the proportion of the total influenza epidemic months in each zone that fell during October to March or April to September from 1964 to 1975. Allowing for the omissions and errors of large-scale reporting, the percentages look close to (1) 90% and 10%; (2) 60% and 40%; (3) 40% and 60%; and (4) 10% and 90%.[4]

The result shows that epidemic influenza is moving south and north across the surface of the globe every year, crossing the equator twice annually around the equinoxes as it follows the winter months. More direct demonstration of this movement can be seen in Figure 8.5, which shows diagrammatically the progress of the great A(H1N1) epidemic of the 1950–51 season across the continent of Africa. It began at around S30° in the Union of South Africa and traveled northward through the south tropical peoples, then through the north tropical

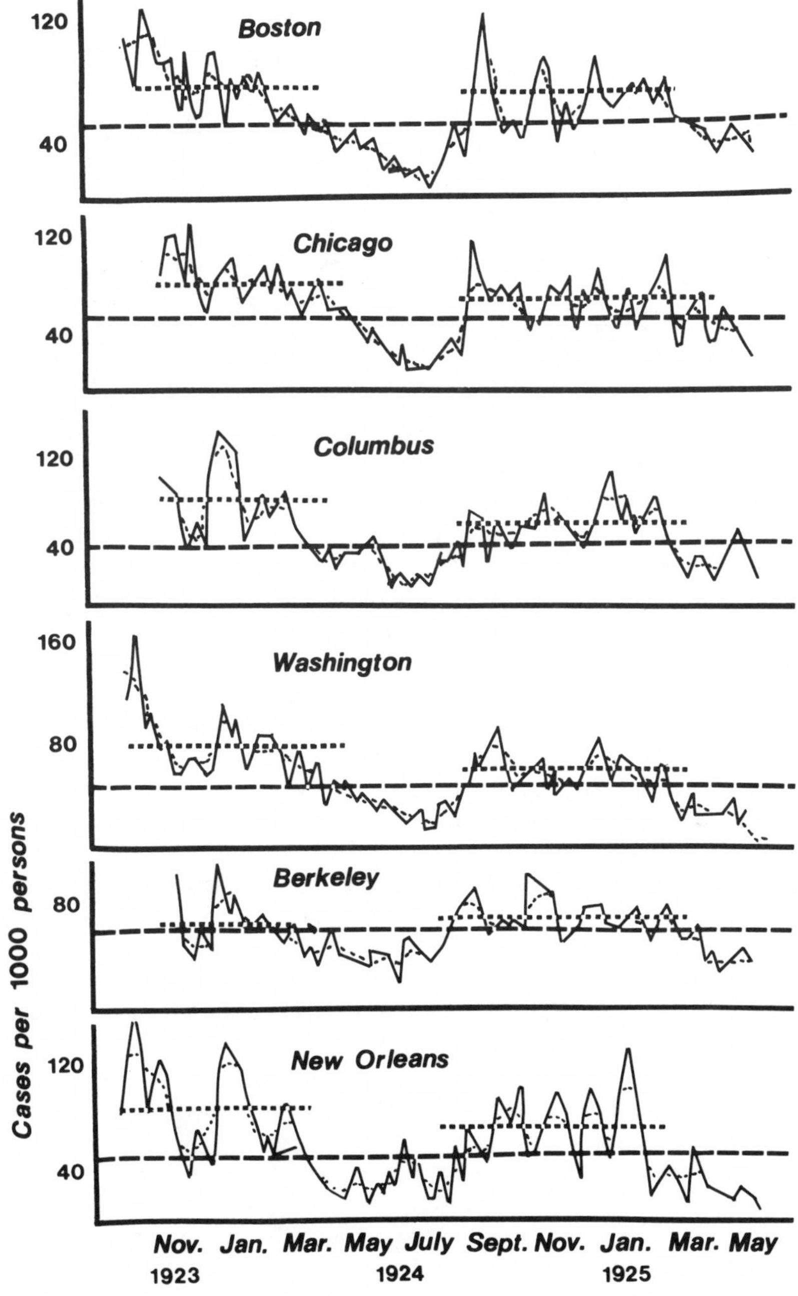

FIGURE 8.3. Colds in students from six campuses. Note seasonal variation, synchronicity, isomorbidity, and endemicity (from Frost and Gover,[3] p. 371, Fig. 2).

TABLE 8.1. Global Epidemic Influenza, 1964–75[a]

Zone (latitudes)	Half-year Oct–Mar	Half-year Apr–Sept	Total
N30°–70°	91.8	8.2	100.0
N 0°–29°	58.9	41.1	100.0
S 0°–29°	35.8	64.2	100.0
S30°–70°	10.5	89.5	100.0

[a]The percentage of the zonal total of epidemic months falling in the half-years October–March and April–September in each of the four major zones of latitude. (From Hope-Simpson,[4] p. 40, Table 3; reproduced with permission from *Epidemiology and Infection*. Data from reports to WHO, Geneva.)

peoples until it reached the people living on the southern shores of the Mediterranean sea some six months later.

The seasonal influence often operates contemporaneously at places lying at the same latitude whatever their longitude. Figure 8.6 shows that influenza epidemics in Cirencester, England (N52°, W2°) occurred at the same time as those affecting the people of Czechoslovakia (N46°–52°, W12°–25°). In addition to the epidemics in the two communities being simultaneous and broadly similar in character, the type of virus, A or B, causing the outbreak corresponded and then, when the strain of influenza A subtype drifted, the changes occurred in both places in the same season. During the same seasons between 1968 and 1974, a similar conformity of influenza epidemics and virus strains occurred in the people of Seattle in the United States, which is situated at around the same northern latitude (N47°–35°) but is even more longitudinally remote from Cirencester (W122.20°).[5]

Contrast this experience with that of two communities living at widely different latitudes antipodeally north and south of the equator, namely, England and Wales (N50°–55°) and New South Wales, Australia (S29°–38°). Figure 8.7 shows the mortality from influenza in both places form 1967 until 1973, a period that included not only epidemics caused by types A and B influenza viruses and antigenic drifts of the A(H3N2) subtype from A/HK/1/68 to A/England/42/72, but also the antigenic shift from the Asian A(H2N2) to the Hong Kong A(H3N2) subtype. Here again the correspondence between the series of epidemics in the two communities was almost identical with the one striking and consistent difference in timing that an interval of approximately six months regularly separated the northern epidemic from its counterpart in the southern hemisphere.

The conclusion seems inescapable that, viewed on a global scale, epidemic influenza is moving annually south and then north through the world population, a smooth yearly scanning of the world very different from the local episodic

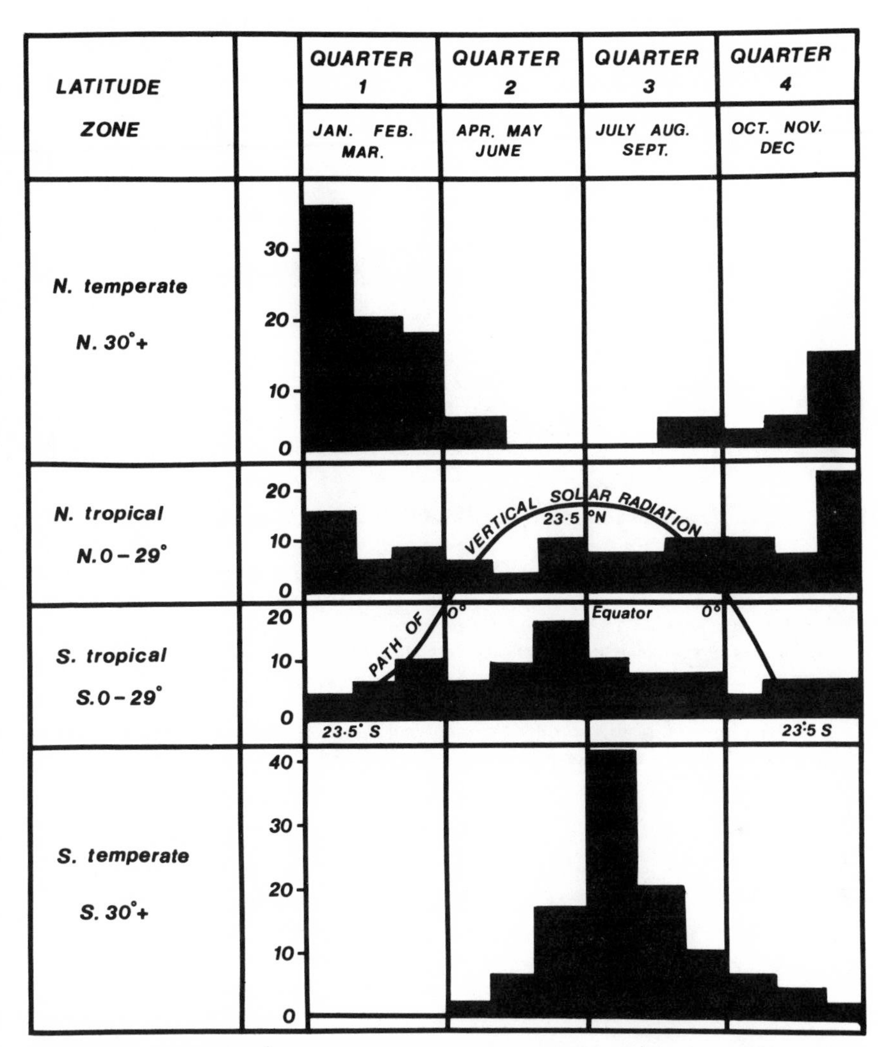

FIGURE 8.4. The distribution of world epidemic influenza 1964 to 1975 in four major zones of latitude. In each zone the percentage of the zone's epidemic months is shown monthly. The epidemic months cluster around local midwinter in both temperature zones whereas in both tropical zones they show a transition, each approaching the distribution in its own temperate zone. See also Table 8.1 (data extracted from the *Weekly Epidemiological Record* of WHO; Hope-Simpson,[4] p. 39, Fig. 2: reproduced with permission from *Epidemiology and Infection*).

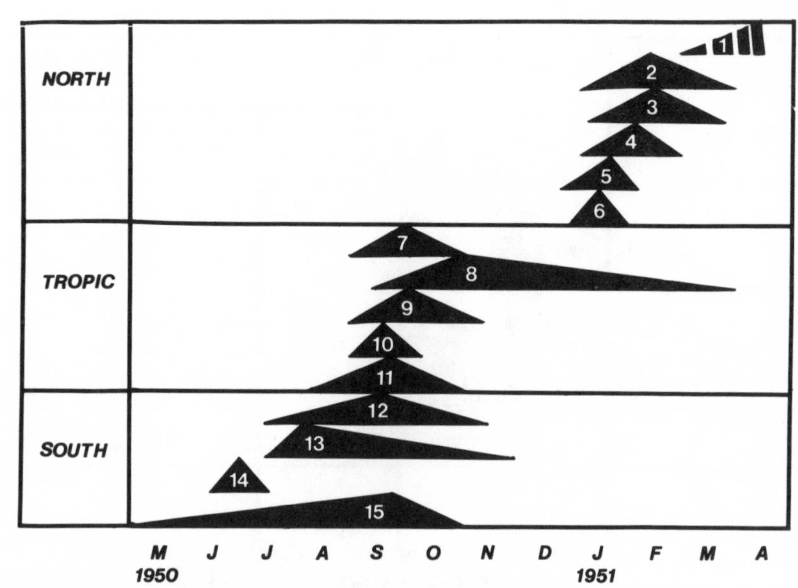

FIGURE 8.5. The 1950–51 influenza epidemic crossing the African continent from South to North. This perhaps illustrates the movement of the proposed seasonally mediated stimulus reactivating persistent virus in ubiquitous carriers and so permitting epidemics to occur in their nonimmune companions. 1. Tunisia, 2. Morocco, 3. Algeria, 4. Tangier, 5. Tripolitania, 6. Egypt, 7. French Equatorial Africa, 8. French West Africa, 9. Nigeria, 10. Cameroons, 11. Spanish Guinea, 12. Basutoland, 13. Nyasaland, 14. Madagascar, 15. Union of South Africa. The figure was constructed in June 1951, 28 years before the new concept, from records received by the influenza department of WHO, Geneva (from Hope-Simpson,[8] p. 172, Fig. 1).

picture of the disease. Neither phenomenon can be explained by the current concept of direct spread. Before considering how the new concept may be able to explain this "transequatorial swing," it is necessary to consider the nature of season and of seasonal phenomena in general.

THE CAUSE OF SEASONS AND SEASONAL PHENOMENA: A NATURAL LAW

Seasons depend on the variations in the complex radiation received by the Earth from the sun. This extraterrestrial agency operates as follows: the Earth performs two regular motions relative to the sun; its daily rotation around its own north–south axis and its annual orbit around the sun. Were the plane of its daily rotation the same as the plane of its annual orbit, there could be no seasons because vertical solar radiation would be falling monotonously on the spinning equator

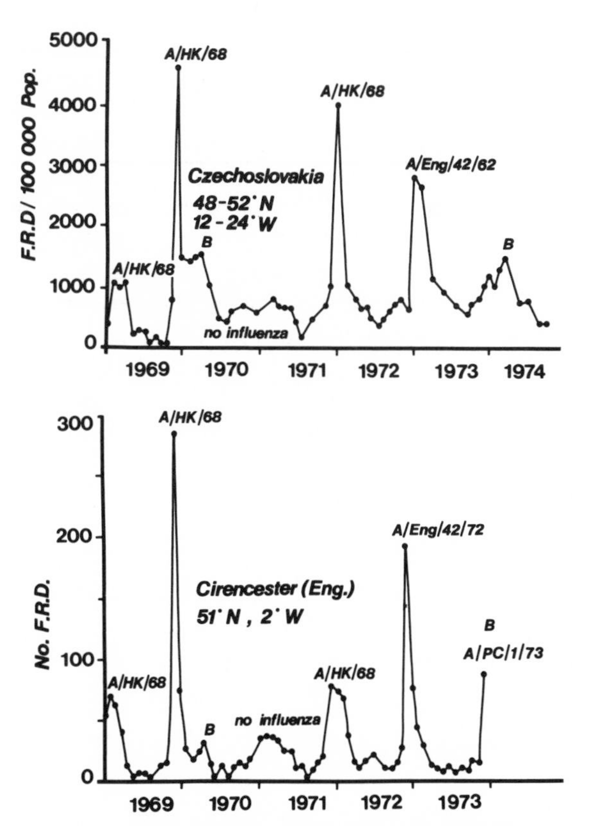

FIGURE 8.6. Influence of latitude upon epidemic influenza. Epidemics 1969–74 at similar latitude but different longitude were contemporaneous and caused by similar strains of the shifting and drifting influenza A virus (from Hope-Simpson,[4] p. 40, Fig. 3; reproduced with permission from *Epidemiology and Infection*).

yearlong. But this is not the case. The plane of daily rotation is tilted about 23.5° in relation to the plane of annual orbit. Vertical solar radiation is compelled thereby to travel a sinuous annual path through the tropical belt from its southern extremity, the Tropic of Capricorn, each 22 December to its northern limit, the Tropic of Cancer, each 21 June and back south again. The midsummer journey of vertical solar radiation crosses the equator twice yearly at the spring and autumn equinoxes. The reader cannot fail to notice that the transequatorial swing of epidemic influenza parallels the path taken by vertical solar radiation on a sort of winter journey about six months later.

The regular variations in the composition, duration, and intensity of the solar radiation falling on different areas of the surface of the globe cause the seasons and all seasonal phenomena. We are here encountering an immutable natural law that seems to allow of no exceptions, namely that: All seasonal phenomena are ultimately caused by the variations in solar radiation resulting from the 23.5° tilt of Earth's rotational plane relative to the plane of its annual orbit.

The consequences are apparent everywhere. Polar icecaps and mountain snowcaps and glaciers expand for half the year and contract for the other half year causing oceans to rise and fall proportionately. Belts of photoperiod and temperature shift north and south each year, and the length of day and night is reciprocally changing to different degrees in different latitudes. The prevalent winds and ocean currents alter their courses seasonally so that local climates have seasonal changes. Such more or less direct physical effects have, in the long course of biological evolution, led to countless adaptations evolved by plants and animals whereby they are able to evade the rigors and take advantage of opportunities that result.

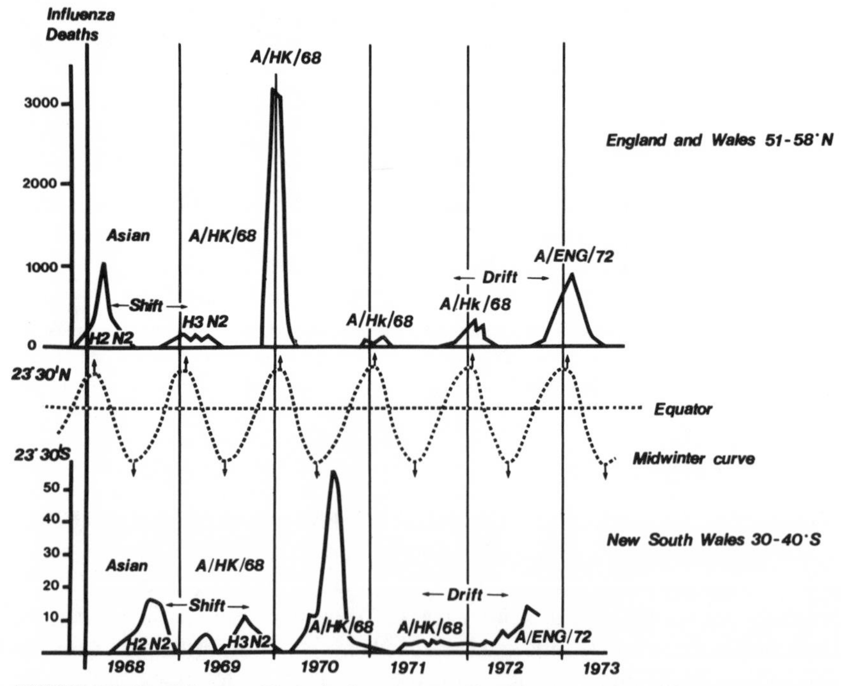

FIGURE 8.7. The influence of latitude. In contrast to Fig. 8.6, populations living at widely different latitudes in different hemispheres regularly experience about six months difference in the timing of the influenza epidemics although otherwise the corresponding epidemics are similar. This figure compares the influenzal mortality in England and Wales with that in New South Wales, Australia from 1968 to 1973. The antigenic shift of influenza A virus from H2 to H3 in 1968–69 and the drift from A/HK/68 to A/Eng/72 appeared punctually in the corresponding epidemic in both hemispheres (from Hope-Simpson,[4] p. 41, Fig. 4; reproduced with permission from *Epidemiology and Infection*).

The harvests of the farmer and the reproduction of many animals and plants are governed by these solar variations and so must move to the pace of the extraterrestrially determined rhythm.

Seasonal diseases cannot escape the law that all seasonal phenomena are ultimately determined by these variations in solar radiation however indirectly the effects are being mediated. The epidemiologist has the task of identifying the chain of intermediate mechanisms through which the prime cause (the variation in solar radiation) is operating its seasonal influence. Sometimes a portion of the chain is evident, as in diseases that are limited to a particular time of year in a limited geographical distribution. For example, human diseases caused by an agent transmitted by a biting arthropod may be limited by the season of the life cycle of the vector. The other parts of the chain, those that are mediating the seasonal life cycle of the arthropod vector, usually remain obscure.

The mechanisms that mediate the influence of variations in solar radiation to cause influenza epidemics seasonally have not yet been identified—indeed, they have scarcely yet been sought—but its seasonal nature must be included as a key feature in any valid concept of influenzal epidemiology.

SEASON AND THE COMMON COLD

Before discussing the seasonal nature of influenza it will be useful to consider that of an even commoner ailment. The seasonal character of the common cold is not in doubt, and it lives up to its name in the peculiarly intimate relationship that it has with the seasonal temperature changes.

The relationship was well shown by Wade Hampton Frost and Mary Gover[3] in the study referred to earlier of the minor respiratory illnesses of groups of volunteering students in university campuses scattered widely across the United States. The study covered 18 months from September 1923 until June 1925 (Fig. 8.3). Their findings show how strongly these illnesses were associated with season and that their exacerbations and diminutions coincided closely on widely separate campuses. There is a striking similarity between the complex variations in all the morbidity curves.

The paper by Frost and Gover did not appear until September 1932 so that they were able to mention the findings of J.J. van Loghem, Sr,[6] that were first published in the Dutch language in 1928. It was he who first called attention to the remarkable secular behavior of the common cold although his study was performed two years later than that of the American authors. From September 1927 until June 1928, van Loghem conducted a weekly postal canvass of 1500 volunteer families living in different departments of the Netherlands. To his surprise the complicated curves of morbidity of the groups in each department were so similar that one curve might have represented the lot (Fig. 8.8). Not only were the peaks

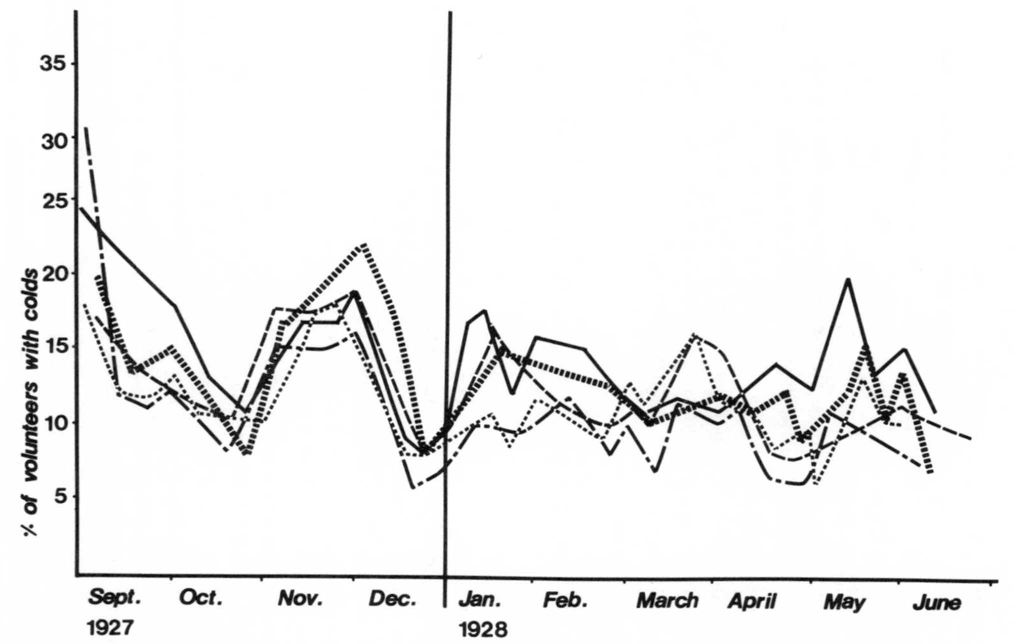

FIGURE 8.8. Colds in the Netherlands. A postal canvass of volunteers in five regions illustrated the phenomena of synchronicity (the morbidity curves are similar and concordant), isomorbidity (of approximately equal prevalence), and endemicity (as opposed to sporadic epidemicity of poliomyelitis or influenza) (from van Loghem[6]).

and troughs nearly *simultaneous* but also the proportion of volunteers attacked in each group was approximately continuously equal (*isomorbidity*). The illnesses were never absent from any of the communities during the 37 weeks of the study (*endemicity*). Figure 8.8 shows the curves of morbidity from Northern Holland, Gröningen, Amsterdam, Utrecht, and Zeeland superposed to emphasize the characteristics of isomorbidity, synchronicity, and endemicity. Frost and Gover realized that the epidemiological problem involved in isomorbidity, which they described as "uniformity of attack rate in widely different places," as indicating that "climate [as opposed to season] is a factor of much less importance than would be supposed." Their figure (8.3) shows all three features emphasized by van Loghem.

From 1954 to 1957 some 370 volunteers, served by the general practice in Cirencester, kept a daily record of the presence or absence of seven common symptoms of acute respiratory diseases.[7] Figure 8.9, based on the morbidity analysis of the first three years and of meteorological records, illustrates the close association between the morbidity from colds and the inversion of the seasonal temperature.[8] Despite inevitable errors in such long-term studies, the inverse concordance between the temperature and morbidity was so close that a decline in one foot earth temperature of one Fahrenheit degree was accompanied by a rise of 1% in the morbidity, and similarly as the temperature rose, so the morbidity decreased *pari passu*. The correlation continued throughout the four years, apart from brief aberrations, some of which were attributable to influenza epidemics.

Measles-type case-to-case transmission cannot by itself explain the epidemiology of the large group of acute respiratory diseases comprehended under the name of the common cold, nor can it by itself explain the seasonal behavior of any other seasonal disease. The discussion on the epidemiology of colds is introduced as an example of the influence of season on a group of respiratory diseases other than influenza. We cannot here consider mechanisms that may be mediating the variations in solar radiation to cause the seasonal character of colds, but they evidently differ from those that mediate the seasonal behavior of influenza. Colds are present yearlong in all communities whereas influenza is almost completely absent except for the few weeks or months when an epidemic is present. Van Loghem contrasted the endemicity of colds with the sporadic epidemic character of acute anterior poliomyelitis in the Netherlands. Influenza would seem also to fit his sporadic epidemic category. It shows no such close association with seasonal temperature changes as characterized the behavior of colds in the Cirencester community.

THE ROLE OF SEASON IN THE EPIDEMIOLOGY OF INFLUENZA

We have emphasized that epidemic influenza is seasonal and that all seasonal phenomena are ultimately caused by variations in solar radiation caused by the tilt

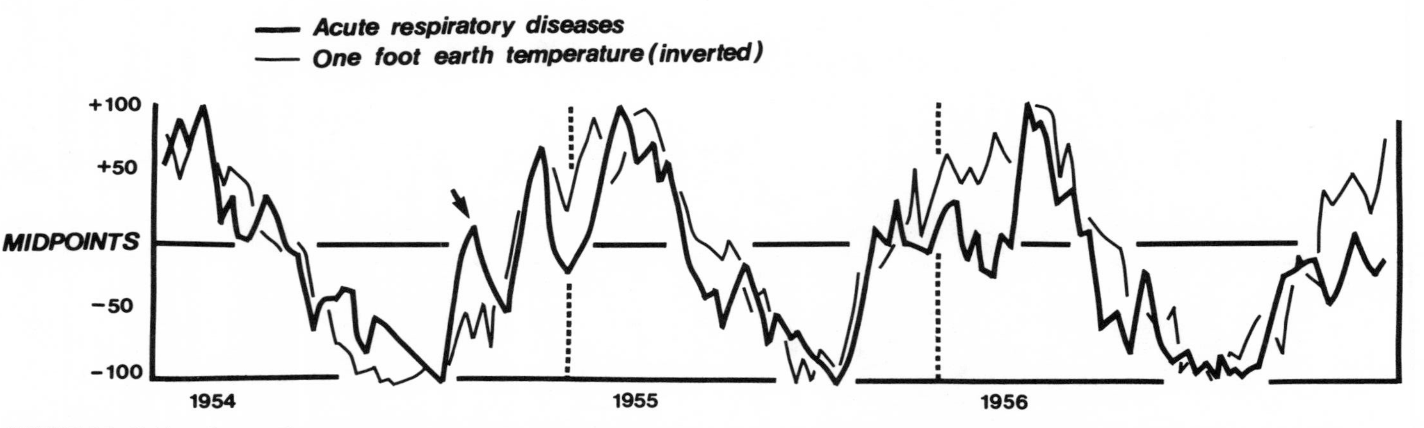

FIGURE 8.9. Colds and seasonal temperature. Weekly percentage of volunteers with acute respiratory diseases above or below the annual mid-value compared with the inversion of the average weekly 1 foot earth temperature similarly calculated. For each 1°F rise in seasonal temperature the morbidity falls by about 1% and vice versa, as the temperature falls so the morbidity rises. Arrow indicates influenza in 1954 (from Hope-Simpson,[7] p. 596, Fig. 2).

of the plane of rotation of the Earth. We therefore cannot evade facing the difficult question: What types of intermediate mechanisms are we looking for that would be capable of mediating the influence of the seasonal variations in solar radiation in a manner that could bring about not only seasonal epidemics of influenza, but also such accompanying phenomena as the antigenic changes in the parasite and the prodigious rapidity of its worldwide geographical spread?

The current concept of direct spread is unable to explain the behavior, and attempts that are made to modify and adapt that concept run into so many difficulties that it seems logical to abandon it in order to assess alternative hypotheses that assume that the virus cannot be transmitted directly from the sick person.

What alternatives are available? The parasite must at some point somehow be transmitted from its human host to reach another person and secure survival of its species. If it is not transmitted during the illness of its host, it must be transmitted either before or after his illness. Both possibilities have been proposed. Evidence against the correctness of the hypothesis that the virus becomes latent before the influenzal illness was given in the last chapter.

The new concept proposes that the virus is being transmitted by the influenzal patient long after he has recovered from the illness, and evidence supporting the proposition will be provided. It suggests that during human influenzal illness the virus enters a mode of noninfectious persistence too rapidly to be transmitted, and the recovered patient becomes a symptomless carrier of noninfectious noncytopathic virus. The seasonally mediated influence must operate by recalling the virus to brief infectiousness in the carrier usually without renewing his illness. His nonimmune companions, however, are then at risk, and if infected promptly develop an attack of influenza, whereupon they in their turn become carriers. The seasonal influence would appear to be mediating its stimulus at a variable time around the winter solstice in the temperate zones. The timing explains why the transequatorial swing of epidemic influenza follows the path across the globe taken by vertical solar radiation about six months earlier.

The new concept, if correct, provides a simple explanation of most of the difficulties besetting the current concept. It explains why influenza epidemics are seasonal and why they may be contemporaneous in widely separated areas at the same latitude. It envisages the world population as being always and almost everywhere seeded with symptomless carriers of noninfectious influenza virus. The seasonally mediated stimulus by which the virus is reactivated in these carriers is traveling continuously south and north across the surface of the globe each year as remorselessly as the vertical solar radiation is traveling north and south. The infected nonimmune companions of carriers are the patients composing the epidemics that are developing in the wake of the annual journey of the seasonal stimulus, the initiation of whose origin is extraterrestrial. Epidemic influenza is thus seen to be a crop, and, as with other crops, some years are good influenza years, and other years produce a poor crop of influenza cases.

SUMMARY OF CHAPTERS 7 AND 8

These two chapters have called attention to the powerlessness of the current concept of direct spread to explain several familiar features in the behavior of epidemic influenza. The virus virtually disappears during the long months between successive epidemics. Epidemics cease in situations in which abundant non-immune subjects are available to support their continuation. Influenza epidemics differ in size and character. some explode over a huge area out of an "epidemiological vacuum," a long period in which no influenza virus has been isolated for many months and no communication can be traced between the earliest cases. Transmission within carefully monitored households is often low even during severe epidemics, so that the apparent intrahousehold attack rate is small. No serial interval can be demonstrated within affected households. The influenzal experience of small communities mirrors that of the whole United Kingdom even when there are frequent antigenic changes in the virus. The age distribution of persons attacked in successive epidemics does not show the expected depression of the average age. All these features are readily explained by the new concept that transmission of the virus is almost entirely limited to carriers of noninfectious persistent virus, who have suffered influenza in a previous season, whose persistent virus colony is seasonally reactivated to infectiousness. The problem of antigenic change is discussed in the next chapters.

REFERENCES

1. Glezen WP, Six HR, Perrotta DM *et al*: Epidemics and their causative viruses—community experience, in Stuart-Harris CH, Potter CW (eds): *The Molecular Virology and Epidemiology of Influenza*. London, Academic Press, 1984.
2. Pringle J: in Thompson T (ed): *Annuals of Influenza in Great Britain*. London, Sydenham Society, 1852, pp 89–90.
3. Frost WH, Gover M: The incidence and time distribution of common colds in several groups kept under continuous observation. *Public Health Rep*, 47, pp 1815–1841, 1932. Also in Maxcy's *Papers of Wade Hampton Frost*. New York: The Commonwealth Fund, 1941, pp 359–392.
4. Hope-Simpson RE: The role of season in the epidemiology of influenza. *J Hyg* (Camb.) 86:35–47, 1981.
5. Foy HM, Cooney MK, Allen I: Longitudinal studies of types A and B influenza among Seattle schoolchildren and families. 1968–1974. *J Infect Dis* 134:362–368, 1976.
6. van Loghem JJ: An epidemiological contribution to the knowledge of the respiratory diseases. *J Hyg* (Camb.) 28:33–54, 1928.

7. Hope-Simpson RE: The epidemiology of non-infectious diseases. (a) Common upper respiratory diseases. *R Soc Health J* 78:593–599, 1958.
8. Hope-Simpson RE: The influence of season upon type A influenza, in Tromp SW, Bouma JJ (eds): *Biometerology: The Impact of Weather and Climate on Animals and Man (period 1973–1978)*. London, Heyden & Son, 1979, p 170–185.
9. Cliff AD, Haggett P, Ord JK: *Spatial Aspects of Influenza Epidemics*. London, Pion, 1986.

9

The Explanation of Antigenic Drift

FEATURES OF ANTIGENIC DRIFT OF INFLUENZA VIRUS

In the last chapter we saw that influenza is seasonal and that its seasonal epidemicity poses insoluble difficulties for the current concept that the virus survives solely by direct transmissions from the sick persons to cause influenza in their infected companions. We therefore tentatively proposed as an alternative concept that the virus so rapidly adopts a noninfectious mode of persistence that it is not normally transmissible during the illness and that the ex-influenza patient becomes a symptomless carrier. Season is in some way involved as an important factor in the epidemiology of influenza, and we suggested that the mechanism whereby season exerts its influence may be by provoking the stimulus that reactivates the virus to infectiousness within these carriers.

This new concept at once overcomes the difficulty of explaining why influenza epidemics are seasonal because the hypothesis has been designed to do so. We must now examine how it fares in the attempt to explain the other seasonal characteristics of influenza, namely, the variations in antigenicity of the virus and the phenomena that are associated with them. Antigenic shift and antigenic drift both occur seasonally and both are associated with the "vanishing trick"—the disappearance of the prevalent predecessor strain(s)—and with the achievement of wide distribution of the successor strain(s) within a single season, worldwide in the case of antigenic shift.

An attack of influenza provokes prolonged immunity against reinfection by strains identical with the causal virus. The antigenic changes caused by the minor mutations of drift are usually insufficient to enable the resulting variant to bypass such protection and cause another attack in the same person. Second attacks of influenza caused by drifted variants of the same major serotype are therefore uncommon. However, as mentioned in Chapter 4, pp. 40–41, in 1946 a larger mutation in the H-coding gene so radically changed the hemagglutinin that the novel strain was able to breach the immunity provided by its related predecessors.

A similar large mutation in the H-coding gene probably occurred in 1918 and another in 1928. The great consequences of these major mutations resembled those of antigenic shift and are accordingly discussed in the next chapter.

Influenza B virus mutates less frequently than type A virus so that it drifts antigenically more slowly. The same strain of influenza B virus often remains prevalent in the human population for many seasons, whereas the master sequence of influenza A virus may be changed in several successive epidemic seasons.

Not uncommonly, more than one minor variant of the same major serotype of influenza virus will co-circulate during the same season (Table 9.1). Each may have its highest prevalence in a different area, but both will be widely distributed and their areas interpenetrate.

The two phenomena that often accompany both drifts and shifts are of great theoretical importance. The first, the *vanishing trick*, is the disappearance of the previously prevalent strain as soon as its successor appears. For example, A/Port Chalmers/1/73 (H3N2) strains, which had caused all the human influenza in a large part of the global surface the 1974–75 season, disappeared in the subsequent season. Few explanations of such illogical behavior have been attempted. Rival serotypes of many other microorganisms co-circulate in the same human community. Types A and B influenza viruses are frequently epidemic at the same time. Two subtypes of type A influenza virus, H3N2 and H1N1, have been co-circulating since 1977, often epidemically present in the same communities. It is perplexing that a minor or a major variant should so often vanish abruptly as soon as its successor appears.

The second phenomenon is almost certainly connected with the first, namely,

TABLE 9.1. The Geographical Distribution of A(H2N2) Influenza Virus Variants in the 1967–68 Epidemic around Cirencester

Locality	A/England/68/68	A/Tokyo/67	Unidentified	Total
Ampney St. Peter	1	—	3	4
Cirencester	16	1	1	18
Poulton	2	—	1	3
Baunton	—	—	2	2
South Cerney	6	—	2	2
Frampton Mansell	1	—	—	1
Coates	2	—	—	2
Stratton	2	—	—	2
Ampney Crucis	4	2	—	6
Kemble	1	—	—	1
Ablington	—	1	—	1
Bibury	—	1	—	1
Total	35	5	7	47

that within a single season the predecessor is replaced by its successor throughout the area of its previous prevalence. In the example quoted above, the A/Port Chalmers/1/73 strain was replaced throughout the enormous area of its 1874–75 prevalence in the 1975–76 season by the A/Victoria/3/75 (H3N2) strain. A strange metamorphosis seemed to have taken place in which the earlier strain had somehow been transmuted into the later. The new hypothesis of antigenic drift given below offers a simple explanation of both phenomena (pp. 101–115, this chapter).

THE CURRENT EXPLANATION OF ANTIGENIC DRIFT

Earlier it was explained that antigenic drift is caused by minor changes in the amino acid sequences at various sites on the hemagglutinin molecules that project as spikes from the surface of the virus. The changes are caused by point mutations in the RNA of the gene that codes for the hemagglutinin. Such mutations occur from time to time on all the genes of the influenza viral chromosome, but only those on the genes coding for the external proteins, hemagglutinin and neuraminidase, cause antigenic drift, hemagglutinin being the more antigenically important. So much is physiological fact. Both surface proteins are strongly antigenic and stimulate the host to produce specific antibodies. The molecular mechanism that causes antigenic drift has been defined as follows: "Antigenic drift, the sequential replacement of human influenza A [and B] viruses by antigenically novel strains, is caused by the interplay of viral mutability and immunological selection."[1] The selective operation of neutralizing antibody can be demonstrated by causing antigenic drift by infecting partially immunized laboratory animals or by cell culture in the presence of homologous antiserum. Webster and Laver suggest that the mechanism also explains the vanishing trick—the disappearance of "outmoded strains" from the population—but do not make the explanation clear. Later, when discussing the vanishing trick in antigenic shift, they candidly admit that it is a baffling conundrum.[1(p. 309)]

The current belief holds that antigenic drift occurs because of "herd immunity," that is, the immune pressure on the virus of a partially immunized human community. On that view it is the recipient who is seen as exerting the proposed immune pressure on the virus. The donor, transmitting the virus from his illness, has not had time to develop his specific immunity. In a partially immunized community the minor variants of the virus are supposed to find themselves at a selective advantage over the previous prevalent strain and so will tend to replace it.

This superficially attractive hypothesis cannot survive closer scrutiny. The "partially immune community" is not a community of partially immune persons, but a community in which some persons have become immune and the others are nonimmune. Second attacks are relatively uncommon, so the recipients who develop influenza and from whom the drifted variant is isolated are persons in the nonimmune portion of the community who could not have exerted the proposed immune selective pressure to cause the antigenic drift.

If the recipients cannot exert the immune pressure and the donor cannot do so during his illness, the current belief is untenable. Several days must elapse after infection before antibody begins to appear in the circulation and the rate of its increase is not very rapid during a primary infection with an influenza virus. Measles virus, which is transmitted during the illness of measles and therefore does not then encounter its own victim's immunity, has remained antigenically stable for decades and possibly for centuries. When measles virus has the opportunity to encounter its specific antibody in persistent infection of cell cultures in the laboratory, antigenic variants appear. They are also found in the rare chronic human measles infection known as subacute sclerosing panencephalitis.

ANDREWES'S HYPOTHESIS OF A DOSE-RELATED RESPONSE

Christopher Andrewes[2] found that by adjusting the dose of influenza virus administered to mice he could obtain a spectrum of response varying from inapparent infection to severe and fatal pneumonia. It seemed to him reasonable to suppose that humans too might be responding differentially according to the dose of the virus they happened to catch from a sick person.

Let us suppose a household in which six nonimmune persons are exposed to an infectious case of influenza; four of them who had only minimal contact with the patient will have received a small dose, whereas the other two, who nursed him, will have sustained a larger dose of influenza virus. Andrewes's hypothesis proposes that the four lightly infected persons may develop an effective but transient immunity without falling ill, whereas the two more heavily infected persons would suffer an attack of influenza and develop a solid, often lifelong, immunity. The transient immunity would be adequate to protect the four symptomless persons from catching influenza from their two sick housemates, but it would have faded before the next influenza season when they would again be at risk of an attack of influenza should they encounter the virus.

The hypothesis is attractive because it accords with both Andrewes's laboratory findings (although he did not say if his subclinically infected mice showed transient immunity) and with the serological finding that the population immunity in human communities rises after an epidemic and declines before the next one. Although Andrewes did not claim it, the hypothesis also offers an explanation for antigenic drift. The symptomless minimally infected people might retain enough of their waning antibody next season to select a mutant in preference to the identical strain that had infected them in the previous season. They would therefore suffer influenza caused by a minor variant of the same subtype, and antigenic drift and the vanishing trick would have occurred in that portion of the community.

Andrewes himself pointed out the serious weakness in his hypothesis. The vanishing trick is startling in its near completeness. The previous strain disappears and is replaced almost completely throughout the whole area, large or small, in which it was prevalent. To produce that result, everybody in the community who did not suffer an attack of influenza must have suffered a symptomless infection so that no nonimmune persons remained, otherwise in a few seasons there would exist a medley of co-prevalent minor variants to a degree that is never encountered.

Were that the only objection, the hypothesis might yet be correct. It might be demonstrating that influenza is so infectious that almost everyone is infected either overtly or asymptomatically in each influenza season. There are, however, other objections. The hypothesis offers no explanation of how the virus is surviving from one epidemic until the next epidemic many months later. Andrewes concedes that it must be adopting some mode of latency. His hypothesis is not compatible with the interepidemic survival of the virus by continuous chains of overt and covert infections. Moreover, it does not explain the seasonal nature of influenza.

Andrewes's hypothesis is not incompatible as an element within the new concept proposed in this book, but it seems not to be necessary. The new concept is able to explain the phenomena of seasonal epidemics, interseasonal absence of the virus, antigenic drift, the vanishing trick, and rapid replacement by the novel strain(s) without the assumptions of the dose-related response.

LABORATORY PRODUCTION OF ANTIGENIC DRIFT OF INFLUENZA VIRUS

During the great influenza A epidemic of the 1950–51 season, Isaacs[3] in London, England and independently Archetti and Horsfall[4] in the United States made a seminal observation. Two minor variants of the A (H1N1) subtype, "Scandinavian" and "Liverpool," were co-circulating in many parts of the world. In both laboratories it was discovered that if the Scandinavian strain were made to infect a fertilized chicken egg in the presence of Scandinavian antibody, the strain harvested a few days later would turn out to be Liverpool. Vice versa, an appropriate dose of homologous antibody induced Liverpool virus-infected eggs to yield a harvest of Scandinavian virus.

The findings have been repeated with other strains of influenza virus in many laboratories throughout the world. St. Groth, first in Burnet's laboratory at the Hall Institute in Melbourne, Australia,[5] and later with Hannoun in Europe,[6] found that he could mimic natural sequential drift of influenza A virus, and he succeeded in producing a prophetic laboratory strain almost identical to one that appeared naturally at a later date. These workers had been hoping that in this way they might have been able to anticipate the natural antigenic variations of the virus, to the great benefit of vaccination against influenza. Unfortunately, the direction taken by the natural sequences of antigenic drift differed from the sequences of minor variants produced in the laboratory. Their other claim to have produced an antigenic shift of influenza A virus subtype could not be repeated.

DRIFT AS EXPLAINED BY THE NEW CONCEPT: AN EXAMPLE FROM GENERAL PRACTICE

The new concept accepts that antigenic drift must result from "the interplay of viral mutability and immunological selection," but addresses the question: At what point in the relationship between the human host and the influenza parasite does the encounter of antigen with antibody occur? Does it occur within the donor who is transmitting the virus or in the recipient whom he is infecting?

It may be helpful to describe the experience of a family living in the Cotswold Hills in Gloucestershire, England, to gain a clear understanding of how the new concept explains antigenic drift (Fig. 9.1).

Félicité, the wife of the vicar of a rural parish, fell ill with influenza at the beginning of December 1974, the first case in a considerable outbreak in the locality. She was credited by her neighbors with having caused the outbreak, but the new concept claims that she could not have done so because the virus causing her illness had too rapidly become noninfectious and persistent in her tissues. Indeed, no other member of the vicarage household developed influenza in the 1974–75 season. The vicar had suffered a severe attack during a previous A/England/42/72 influenza epidemic, but unfortunately no specimen for virological study was collected from him. No other member of the vicarage household suffered an attack of influenza in that earlier epidemic.

It is suggested that some 13 months after Félicité's influenza the virus persisting in her tissues was reactivated by the seasonally mediated stimulus, and in fact her youngest child, Benedicta, aged 13 years, fell ill on New Year's Day, 1976, with the first case of influenza of the next epidemic in the locality. Next day, Félicité's son William, aged 22, came down with the disease, followed a few days later by her mother Ellen, aged 78. All three were living in the vicarage.

So much for the bare outline of the story. The details are of great interest. The strain of A (H3N2) influenza virus isolated from Félicité on 2 December 1974 was identified as A/Port Chalmers/1/73. This Port Chalmers strain had been responsible for the influenza A that had occurred in most parts of the world during that 1974–75 season. The new concept suggests that by December 1975, Félicité must have been comparable to one of Isaacs's chicken eggs. She had been infected with Port Chalmers strain in December 1974 to which she had responded within a

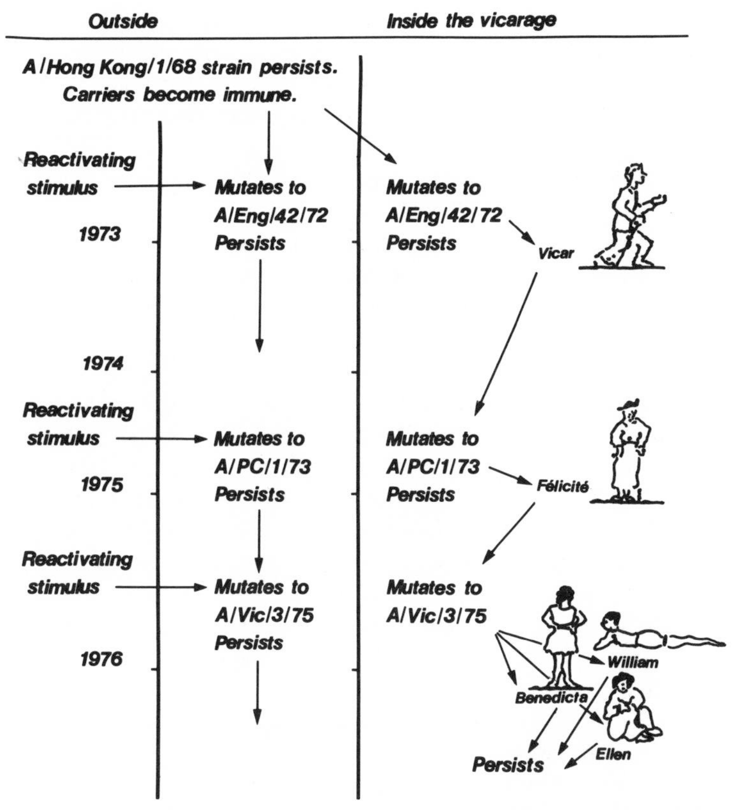

FIGURE 9.1. Antigenic drift of influenza A virus within a vicarage. During the A/England/42/72 (H3N2) influenza epidemic in 1972–73 the vicar was severely attacked. In 1974–75 his wife suffered the first local case during the A/Port Chalmers/1/73 (H3N2) epidemic, again the only case in the household. The remaining members suffered the first cases in the A/Victoria/3/75 (H3N2) epidemic of 1975–76. Similar sequences were occurring over much of the world during 1972-76. The new concept proposes that none could transmit the virus during their illness, because it so rapidly became dormant until reactivated much later by a seasonally mediated stimulus. They were then immune to the parent strain and transmitted a mutant (from Hope-Simpson,[10] p. 78, Fig. 5; reproduced with permission from *PHLS Microbiology Digest*).

fortnight by producing a specific immune reaction. We must picture her next season in December 1975 as carrying in her tissues the noninfectious virus in a mode of persistent innocuous infection and in her blood antibody specific against the Port Chalmers virus. Then came the stimulus that reactivated her virus to become fully infectious. Many of the reconstituted virions will have been identical with the parent Port Chalmers strain. These will at once have been neutralized by the Port Chalmers antibody. In every attack of influenza, although replication produces a vast majority of particles identical with the infecting strain, an assortment of mutants is also produced and it is some of these that evade the immune reaction in the carrier and are transmitted. It was at this point that "the interplay of viral mutability and immunological selection" took place. The "antibody pressure" in the carrier (Félicité) favored the mutants, and her nonimmune companions "selected" from the mutants that had escaped her immunity the one that was "fittest" in an evolutionary sense to survive and continue the influenza A virus species. We were watching evolution taking place before our eyes.

The fittest of these variants of A/Port Chalmers/1/73 strain in that 1975–76 season happened to be one called A/Victoria/3/75, and this Victoria strain was isolated from Benedicta, William, and Ellen. They had harvested it from Félicité just as Isaacs had harvested the Liverpool strain from eggs infected with Scandinavian virus that had been grown in the presence of Scandinavian antibody.

But that is not the end of the story. The millions of persons throughout the world who had been infected with A/Port Chalmers/1/73 strain in the 1974–75 season had suffered the same antigenic input and had responded with a similar immunity. It is the similar antibody response by which the causal organism is commonly identified in an epidemic. The same antigen and the same immune response would probably have resulted in a similar assortment of mutant variants to be shed by the patients and offered to their nonimmune companions who would also have selected the fittest mutant, namely, A/Victoria/3/75. It is a matter of history that in the 1975–76 season A/Port Chalmers/1/73 strains disappeared and were everywhere replaced by A/Victoria/3/75 strains.

In this way the new concept offers a simple explanation not only of antigenic drift, but also of the vanishing trick and of the means whereby the novel strain(s) replace the predecessor(s) in a single season, however wide the area of prevalence. It also explains why these phenomena are all seasonal.

Sometimes more than one mutant of almost equal evolutionary potential may be competing against each other during the reactivation within the carrier. In that case some of the nonimmune companions may select one of the mutants and others may select another. For example, during the 1967–68 A(H2N2) epidemic illustrated by Table 9.1, three cases of influenza occurred in one family in Ampney Crucis village. Christopher, aged 3, and Colin, aged 8, were sharing the same bed and their sister Julie, aged 7, was also ill with influenza in a different bed in the same room. On 25 January, A/England/68/68 strain was isolated from Colin. The next day, A/England/68/68 was isolated from Julie and A/Tokyo/67 strain from Christopher. The new concept envisages the two variants as mutants of almost equal potential that the three children had picked up from the same donor.

During an influenza B epidemic in Cirencester in the winter of 1973–74, twin boys, aged 9, shared the same bed. Specimens taken on 28 December 1973 grew a different minor variant from each boy.

SERIOUS DIFFICULTIES IN EXPLAINING ANTIGENIC DRIFT

A major difficulty faces all hypotheses of antigenic drift if convalescent serum from influenza patients is able to neutralize not only the infecting (parent) strain of the virus, but also its mutant variants. It is necessary to recapitulate briefly the main facts that are known about the mechanism of drift in order to set the scene for a discussion of this vital matter.

Ribonucleic acid, the genetic material of the influenza virion, is inherently mutable, and mutations in the genes coding for the surface proteins hemagglutinin (H) and neuraminidase (N) alter their amino acid sequences. When such alterations occur at one or more of the four or five antigenic sites, *epitopes,* on the H or N molecule, corresponding changes are provoked in the host's antibody response to the infecting virus. Variants produced by mutation at these sites are said to have drifted antigenically from the parent strain, but the term antigenic drift is usually employed when, as often happens, the prevalent parent strain is superseded in the community by one or more of its mutant progeny. The problem that faces us is to explain how this phenomenon can be brought about.

We saw in the section on the explanation of antigenic drift (p. 97) that the current explanation maintains that herd immunity, the immunity of a partially immune community, exerts selective pressure on the transmitted virus that favors the mutant in competition with its parent strain when patients suffering from influenza are shedding the virus. Against this explanation it was pointed out that it is usually the nonimmune companions of the sick who are catching the disease, the very people who cannot be exerting immune selective pressure.

Agreeing that immune selective pressure is important in the mechanism of drift, the new concept evades the above difficulty by proposing a different time for the encounter of the virus with its specific antibody. It suggests that the patient usually cannot transmit the influenza virus during his illness and remains a carrier of the virus in some noninfectious mode until it is seasonally reactivated months later when the carrier has long developed his immunity to it. Thus the virus encounters the immune pressure before, not after, it is transmitted. It is the donor, not the recipient, who exerts the selective pressure.

At this point we encounter major problems. The first concerns the infrequency of mutation. It appears that mutation at a single epitope on the H molecule occurs only at a rate of 1 : 100,000, while simultaneous mutations at more than one epitope occur very rarely indeed. How can such a rare event so often succeed in replacing the prevalent strain in a few months over enormous areas, even worldwide? The problem looks to be insoluble for the current explanation of drift, which sees the donors as shedding the virus during their illness before they have had time to develop the specific antibody. Their nonimmune companions will be receiving at least 100,000-fold more numerous particles of the parent virus than of the mutant strain.

According to the new concept, however, when transmission occurs the selection in favor of the mutant has already been made within immune carriers who are already widely distributed throughout the community.

The second problem appears more intractable. The mutant variants, being closely related to the parent virus, appear to be readily neutralized by the immunity it provokes. Persons seldom suffer a second attack caused by a drift variant of the strain that caused their influenza. Few of the patients in the Asian epidemic of H2N2 influenza A suffered a further attack from that subtype during its 11-year era of prevalence. Even during the last long epidemic during the early months of 1968 when the prevalent strains had drifted antigenically a long way from their 1957 prototype, second attacks caused by the A(H2N2) strains were rare.

There was experimental support for the opinion that drift mutants are neutralized by serum provoked by their parent strain:

> . . . polyclonal animal sera neutralized the antigenic variants to the same extent as parental virus suggesting that new antigenic variants would be easily neutralized in nature. . . . In particular, since an immune individual might be expected to produce antibodies against all four of the antigenic sites on the HA molecule it has been particularly difficult previously to understand how such variants could arise in nature at such low frequencies and subsequently could have escaped neutralization and [could] have spread in the community.[7]

Yet, throughout many years during which recognition has been possible, antigenic drift has been recognized season after season. Some way of escaping neutralization and of achieving wide distribution must therefore be available to these scarce mutants. A technique was needed whereby mutant strains, individually selected, could be tested against immune sera raised against their parent strain.

The appropriate technique became available when Koprowski, Gerhard, and Croce[8] obtained monoclonal antibodies directed against single epitopes on the H and N molecules of influenza viruses. Their results made it possible for Natali, Oxford, and Schild[7] to perform the crucial experiments " . . . to determine whether antigenically variant influenza A viruses selected *in vitro* using monoclonal antibodies to HA were changed, compared with parental virus, in their ability to react with human sera."

Having examined sera from 210 children aged 1–5 years and 197 adults aged 18–32 years, Natali and co-workers were able to make the following observations:

> . . . a proportion of human antisera is able to distinguish between antigenic variants, selected *in vitro,* after passage in a single monoclonal antibody preparation and presumably therefore having an antigenic change in a single epitope, and the parental virions. It would appear that . . . a proportion of human sera, and particularly children's sera, possess a more limited anti-influenza virus antibody repertoire, which is restricted to one or perhaps two non-overlapping epitopes, thus potentially allowing such antigenic variants to escape neutralization and [to] spread in the community.
>
> We can hypothesize therefore that comparable novel antigenic virus mutants could be selected in nature and spread, particularly in young children. Children frequently experience the

highest attack rates during influenza epidemics and may be a primary source of virus spread in the home and community.[7]

The authors then cite evidence that young children produce antibody that is highly specific for the hemagglutinin of their infecting influenza A virus and which does not react with other variants of the same antigenic subtype. Readers will be reminded of the serological findings discussed in Chapter 6 in the section entitled "The Doctrine of Original Sin and Other Unexpected Findings."

These careful experiments have identified the escape routes for antigenic variants, but it is difficult to understand their claim that the findings also explain the epidemic spread of the variants from the sick persons who have not yet become immune. According to their concept the escaping variant must still be facing odds of 1 : 100,000 or more against success in infecting each nonimmune companion of the sick patient because it is still competing against its unneutralized parent strain.

Natali and her colleagues have, however, provided crucial evidence for the hypothesis of antigenic drift advanced by the new concept. The patient, having become a carrier of persistent influenza virus, which becomes infectious under a seasonally mediated stimulus after he has become specifically immune, will neutralize the parent strain and, in most cases, will also neutralize the mutant strains. But these authors have demonstrated that, in some such carriers, the mutant strains alone would escape a more limited antibody repertoire, and only such people qualify as potential donors who can spread the virus. They are widely distributed in the community and so are well placed to procure antigenic drift in a single season over the whole area previously affected by the parent strain.

As indicated, children would be an important source of spread, and evidently some young adults also qualify as potential spreaders. Information concerning older persons in whom immune potential is perhaps waning would be particularly interesting.

Much more evidence is needed; nevertheless, the findings of this study encourage the speculation that a person's first-ever infection with influenza A virus may cause him to be a carrier of persistent virus for a year or two.

THE BEHAVIOR OF NATURAL KILLER LYMPHOCYTES IN INFLUENZAL INFECTIONS

A recent paper that calls for consideration is by Dr. Trushinskaya and the late Professor Zhdanov.[9] It describes an investigation into the anomalous behavior of natural killer (NK) lymphocytes in influenzal infections that casts a new light on part of the mechanism of antigenic drift that may be relevant to the drift hypothesis of the new concept.

These NK cells appear very early, within a few hours of the infection. Other subpopulations of lymphocytes—cytotoxic immune T killer cells, antibody-promoting and -producing B-lymphocytes, B-lymphocyte memory cells—tend to be generated later, five to seven days after infection.

In most viral infections, both the whole virions and their free proteins that accumulate during replication activate the NK cells. Influenzal infection is peculiar in that although the intact influenza virions activate NK cells, their free proteins (hemagglutinin, neuraminidaase, matrix protein, nucleoprotein) inhibit them.

This peculiarity of influenzal infection was found to have an important differential effect on the heterogeneous population of influenza virions. The effect of the earlier population of influenza virions differed from that on the later virions when free proteins had become abundant. The authors regarded this as being the basis of influenzal antigenic drift.

They also found that the immunogenic property of the epitopes on the H molecule of the influenza virion behaves independently of its antigenic property during antigenic drift. The epitopes on the H molecules of earlier A(H1N1) strains lose their *immunogenicity* when they become components of the hemagglutinin of later strains, the drift variants, although they retain their *antigenicity* and are able to bind with antibodies to closely related strains. Conversely, nonimmunogenic determinants of the H molecule antigenic sites become immunogenic when they become components of later drift variants. "On this trait," say Trushinskaya and Zhdanov, "we have attempted to map the H molecule from the viewpoint of this distribution of immunogenic and non-immunogenic determinants for influenza A(H1N1) strains isolated in two periods: 1947–1953 and 1977–1979."

The scheme of the epidemiology of human influenza that they have produced is one of daunting complexity. Cell-mediated immunity is inherently complex and yet they point out that for the sake of simplicity they have omitted several important influences.

They propose that influenzal infection is able to take one or more of 11 different directions simultaneously or successively. They claim proof, direct or indirect, for nine of them (see Fig. 1 of their paper).

The earliest virus population (A) consists during the first two or three days of a number of strains in different proportions, one strain being dominant. The differential action of the NK cells ensures that a quite different population of strains (B) replaces population A during convalescence. A strain that was in a minority in population A has been favored by the differential action and becomes the dominant strain of population B.

The authors make the comment that population B can only be discovered at the beginning of the next epidemic. One is struck by how close their concept is approaching the new concept advanced in this book. They are aware of the similarity and at this point they mention it. Their detailed study of influenzal cell-mediated immunity has suggested a mechanism whereby antigenic drift can be brought about in the body of the human host infected by influenza before the virus is transmitted to other members of the community.

They have, however, run into difficulty by adhering to the current concept that the virus is spreading directly from the influenzal patient during his illness. They have said, for example, that antigenic drift of influenza virus hemagglutinin is based on the immunoselection of spontaneous mutants of the virus *when it is circulating in the human population.* And they are surely mistaken when they add that:

> . . . the specific immunity which developed in a human population after an encounter with one drift variant, does not provide protection from another drift variant which carries insignificant changes in the antigenic determinants, due to substitution of individual aminoacids located at 4–5 sites on the H-molecule.

As we saw in the last section, this statement is contrary to clinical experience in human epidemic influenza, so much so that it required the work of Natali and her colleagues to discover the exceptional persons of whom the statement was correct. Trushinskaya and Zhadanov confess that they are aware of some difficulties that arise from their assumption:

> This viewpoint is now well-known, although many questions remain unanswered. In particular, scientists have for a long time asked why the immune-press [*sic*] does not act on measles and poliomyelitis even though they are just as ubiquitous as influenza.

The immunoselection that causes drift should not be sought in the immune status of the general population because persons who have suffered an attack of influenza are seldom subsequently attacked by influenza caused by a related variant. The authors have provided a valuable description of a mechanisms that accords well with the hypothesis of drift proposed by the new concept, namely, that the human carrier, recovered from influenza, himself provides the immune pressure that selects the variant strain. The proposed seasonal reactivation from noninfectious persistence explains why the new variant is difficult to isolate during convalescence but readily isolated at the beginning of the next epidemic.

The new concept also explains their problem about why measles virus remains antigenically stable in contrast to the variability of influenza virus. Measles virus is transmitted from the measles patient while he is sick, before he has developed his immunity to the virus. The virus therefore does not encounter the specific immune pressure that would necessitate drift as a means of viral survival. If, as proposed here, influenza virus is transmitted long after the influenzal illness, it needs to evade the immunity that it has engendered in its human host. The drift variant would have an advantage over the parental strain in those carriers with a limited antibody repertoire against influenza virus, discovered by the experiments of Natali and her colleagues. The difference between the mutability of influenza virus and the stability of measles virus would thus be readily explained.

Possibly, when more than one variant succeeds in bypassing the immunity engendered by the parent strain, the recipient nonimmune portion of the population may exert a selective function in choosing between the competing variants.

DURATION AND FATE OF PERSISTENT VIRUS IN HUMAN CARRIERS

Persistent infection of cell cultures by influenza virus has not been sustained for much more than 18 months in the laboratory. A similarly short duration of the persistent mode postulated as occurring in human carriers may explain the orderly succession of antigenic drifts and the small number of minor variants that dominate human epidemic influenza each season.

If the persistent state endures only for a year or two in human carriers, what happens to the colonies of the virus at the end of that time? The question is discussed in the next chapter dealing with the problems posed by antigenic shift.

SUMMARY

The current explanation of antigenic drift as caused by herd immunity cannot be sustained because it is not the immune persons in the herd that are infected, but the nonimmune. Moreover, the current hypothesis of direct transmission offers no explanation of the phenomena that often accompany drift, namely, disappearance of the previously prevalent strain and its replacement within a single season by its successor.

The new concept explains antigenic drift by immune pressure within the carrier in whom persistent virus has been seasonally reactivated long after his attack of influenza. The parent strain is neutralized by his immunity so he transmits only its mutants from which his nonimmune companions select the fittest. The previous strain thus automatically vanishes and is replaced next season throughout the area of its prevalence by the most fit mutant. The successful transmitters are carriers whose antibody repertoire to the parent influenza virus is unusually narrow.

REFERENCES

1. Webster RG, Laver WG: Mechanism of antigenic drift, in Kilbourne ED (ed): *The Influenza Viruses and Influenza.* New York, Academic Press, 1975, pp 287, 309.
2. Andrewes CH: Epidemiology of influenza. *J Infect Dis* 128:361–386, 1973.
3. Isaacs A: The 1951 influenza virus. *Proc R Soc Med* 44:801–803, 1951.
4. Archetti I, Horsfall FL: Persistent antigenic variation of Influenza A viruses after incomplete neutralization *in ovo* with heterologous immune serum. *J Exp Med* 92:441–462, 1950.

5. de St. Groth SF: Antigenic variation of influenza viruses. *Arb Paul Ehrlich Inst Bundesant Sera Impfstoffe Frankf* 77:21–34, 1977.
6. de St. Groth SF, Hannoun C: Selection of spontaneous antigenic mutants of influenza A virus (Hong Kong). *CR Acad Sci* (III), 276:1917–1920, 1973.
7. Natali A, Oxford JS, Schild GC: The frequency of naturally occurring antibody to influenza virus antigenic variants selected *in vitro* with monoclonal antibody. *J Hyg* (Camb.) 87:185–190, 1981.
8. Koprowski H, Gerhard W, Croce CM: Production of antibodies against influenza virus by somatic cell hybrids between mouse myeloma and primed spleen cells. *Proc Natl Acad Sci USA* 74:2985, 1977.
9. Trushinskaya GN, Zhdanov VM: The role of natural cytotoxic lymphocytes (natural killers) in the pathogenesis of influenza. *Vopr Virusol* 1:103–110, 1988.
10. Hope-Simpson RE: Simple lessons from research in general practice. *PHLS Microbiol Dig* 7:74–79, 1990.

10

Antigenic Shift of Influenza A Virus

THE CHANGING DEFINITION OF ANTIGENIC SHIFT

A major change in the antigenicity of influenza A virus in 1946 presented a problem to the scientists who were studying the virus. Since the discovery of the virus in 1933, they had become familiar with the almost seasonal antigenic drifts that changed the virus so little that the resulting minor variants had seldom caused a further attack of influenza in persons who had previously been attacked by influenza A virus. They had correctly inferred that antigenic drift was caused by point mutations in the genes coding for one or both of the surface proteins hemagglutinin and neuraminidase.

The change that occurred in 1946 was so much greater than any of the known antigenic drifts that persons who had previously suffered an attack of influenza A were no longer protected against the novel strain. Serological studies showed that the hemagglutinin differed considerable from that of the predecessors, which were later allotted to a subtype called A(H0N1). The change was rated as an antigenic shift to distinguish it from the previous drifts and it was accompanied by the two remarkable phenomena that have already been mentioned. First, as soon as the 1946 strains appeared, the A(H0N1) strains that had caused all the influenza A since 1933 disappeared, and second, the strains, belonging to the new subtype, later called A(H1N1), replaced them worldwide within a single season and continued as virtually the sole cause of influenza A for the next 11 years.

An important question concerned the nature of the 1946 antigenic variation. Was it a matter of degree, a mutation in the H-coding gene larger than those causing antigenic drift? Alternatively, was it caused by an altogether different mechanism? The decision remained doubtful but the majority opinion favored mutation.

The era of world prevalence of A(H1N1) influenza virus came to an abrupt end in 1957 when it was replaced worldwide by another major variant in which both the hemagglutinin and the neuraminidase were found to be much altered. The change was again classed as an antigenic shift and strains of the new subtype were named A(H2N2). They caused almost all the influenza A until 1968 when another antigenic shift affecting the hemagglutinin replaced the A(H2N2) strains by strains of A(H3N2) subtype.

Despite dissentient voices, all three antigenic shifts were at first generally considered to have been caused by mutations on the genes coding for the external proteins until, in the 1970s, it was discovered that this had been true only of the first one. In 1946, the hemagglutinin had indeed been changed by a major mutation in the H-coding gene, whereas in 1957 and in 1968 the mechanism of antigenic shift had not involved a mutation but a reassortment of whole genes.

Genetic reassortment, such as that of 1957 or 1968, occurs when two different influenza A viruses co-infect the same host cell. The eight RNA segments of the viral chromosome are only loosely connected and within the host cell they become separated in order to be individually replicated. They must be reassembled and packaged in correct order to produce viable genomes.

When two different influenza A viruses simultaneously infect a host, numerous cells are co-infected and during viral replication will possess two sets of eight separated viral RNA segments. RNA segments from the different parent viruses are readily exchanged during reassembly in such doubly infected cells, so that the progeny consists not only of strains identical with each parent but also of hybrids containing some genes from one parent and the rest from the other parent. An exchange of genes coding for the internal proteins is not obvious serologically, but an exchange of H- or N-coding genes or both has a profound antigenic effect.

It was found that a great change had been effected in 1957. Only four RNA segments had been conserved from the A(H1N1) subtype strains, and the four novel segments included segment 4 (containing the gene coding for hemagglutinin) and segment 6 (containing the gene coding for neuraminidase). In 1968, only segment 4 had been exchanged.

In 1980 an international committee of experts[1] decided that it was necessary to distinguish the two sorts of antigenic variation, mutation and genetic reassortment, by reclassifying influenza A viruses. The term *antigenic shift* was thereafter to be reserved for the changes resulting from genetic reassortment involving at least the H-coding gene. The term *subtype* was to denote the family of minor variants initiated by an antigenic shift and appearing sequentially by antigenic drifts caused by mutations in the H- and N-coding genes.

The major mutation from H0 to H1 of 1946 was reclassified as an antigenic drift in the new subtype A(H1N1), which also includes the major serotype detected by retrospective serology previously named A(Hswine1N1-like). The classification brings human influenza A viruses into line with the classification of isolates from nonhuman hosts, but it conceals important features of influenzal epidemiology. When discussing the eras of prevalence of the three major serotypes of A(H1N1) subtype, it is necessary to distinguish each by its old name, for example, A(H0N1) and A(Hswine1N1-like). The third needs to be characterized as A(H1N1 old style).

THE ERAS OF PREVALENCE OF MAJOR INFLUENZA A SEROTYPES DURING THE LAST 100 YEARS

Figure 10.1 illustrates the eras during which the major serotypes of influenza A virus have been prevalent in the world population during the 100 years that preceded the writing of this book. The information concerning the first half of the period from about the date of the pandemic of so-called "Russian influenza" in 1889 until the discovery of influenza A virus in 1933 is based on retrospec-

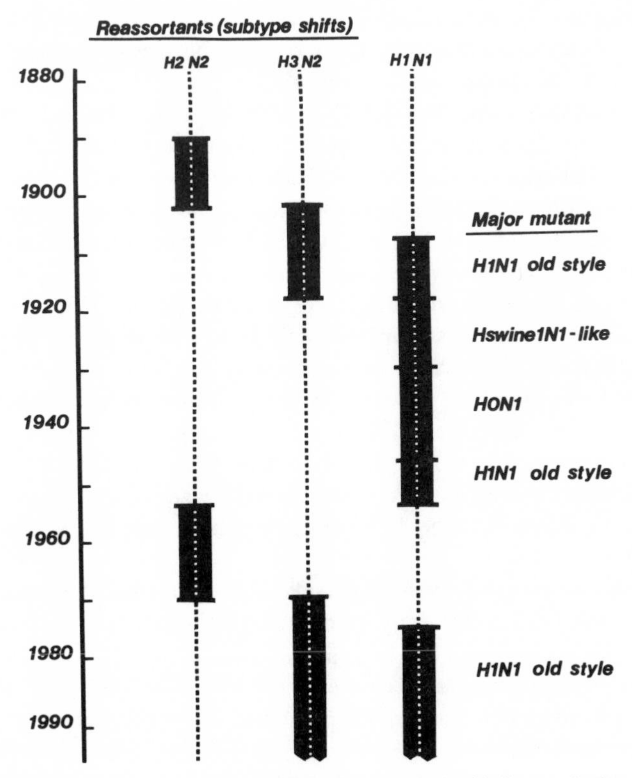

FIGURE 10.1. Eras of prevalence of the major human serotypes of influenza A virus in mankind during the last 100 years.

tive serology and it is therefore, perhaps, less reliable than the information in the second half which is based on molecular virological studies in addition to serology.

The illustration shows that strains of A(H2N2) subtype have had two eras of world prevalence, 1889 to 1900 and 1957 to 1968. A(H3N2) subtype strains have also had two eras, first succeeding the A(H2N2) strains from 1900 until 1918 and then again replacing them from 1968 until the time of writing 1989. Strains of A(H1N1) subtype as currently defined have also had two eras of world prevalence during these 100 years. The first lasted from about 1908 until they were replaced by A(H2N2) strains in 1957 and the second began in 1977 and persists at the time of writing.

The first era of prevalence of A(H1N1) subtype lasted for 50 years, far longer than the prevalences of the other two subtypes, and the era was antigenically complex. Figure 10.1 shows that its major antigenic changes, now classed as drifts, caused a succession of eras of world prevalence of major serotypes of the subtype each of which resembled an era of subtype prevalence except that the successor was a mutant, not a genetic reassortant. They are included in the diagram because their eras of prevalence present epidemiological puzzles analogous to those of antigenic shift and subtype prevalences.

The illustration shows that A(H1N1 old style) strains were co-prevalent with A(H3N2) strains from about 1908 until 1918 when both major serotypes were replaced by A(Hswine1N1-like) strains. The swine-virus-like H1N1 strains caused all the influenza A until 1929, a period sometimes known as the "swine era." They were then replaced by A(H0N1) strains until 1946 when these were replaced by a return of A(H1N1 old style) strains whose 11-year era of world prevalence terminated this first long era of A(H1N1) subtype strains. However, it was as A(H1N1 old style) that the subtype returned in 1977 for its second co-prevalence with A(H3N2) strains, and they are both still with us in 1989.

The bald details of the major serotype prevalences as related above give an inadequate picture of their eras. The reader should bear in mind that the shift in 1889 had inaugurated an era of A(H2N2) prevalence characterized by a series of annual epidemics of varying size until by 1900 it had immunized practically all the people in the world who had not already been immune to the A(H2N2) subtype.

The shift to A(H3N2) strains that replaced them in 1900 also initiated annual outbreaks until 1917 or the early months of 1918 when they in their turn had immunized the people who had been nonimmune to A(H3N2) strains. They had been joined in 1908 for an era of co-prevalence by A(H1N1 old style) strains until by 1918 they too had immunized that portion of the world population previously nonimmune to them.

The historic year, 1918, that witnessed the eclipse of the A(H3N2) subtype and of the A(H1N1 old style) major serotype saw them both replaced worldwide by another major H1 mutant. This antigenic change within the A(H1N1) subtype promptly caused the greatest recorded influenza pandemic. Although no longer regarded as an antigenic shift, the drift that produced the H-swine variant caused consequences similar to those that characterize a shift. The predecessor A(H1N1 old style) strains vanished and the mutant swine-like strain achieved world distribution in the 1918–19 season and continued as the sole cause of influenza A in a series of annual outbreaks until the previously nonimmune persons had become immunized against it throughout the world.

It is believed that the human A(Hswine1N1) strain first infected domestic swine about October 1918 and became established as the cause of swine influenza. It disappeared from mankind as an epidemic strain in 1929 but has remained as a bane of pig farms in many parts of the world. We shall be relating how a return of this virus to a human community caused great alarm in 1976 (Chapter 13: The Fort Dix Influenza Epidemic).

The influenza A story continues with the swine virus being replaced by A(H0N1) strains to cause all the influenza A from 1929 until 1946. Here again one encounters the shift-like result of a major mutation with vanishing of the predecessor strains and replacement worldwide by the successor within a single season. The phenomenon is repeated in 1946 with the return of A(H1N1 old style) after 28 years absence for another 11 years of world prevalence, this time as the sole agent causing type A influenza until 1957.

Then we come to the second A(H2N2) era from 1957 to 1968, followed, as in 1900, by the repetition of the A(H3N2) era from 1968 onward, joined in 1977, as in 1908, by strains of A(H1N1 old style) for their third era of prevalence.

During each of these eras of prevalence the dominant strain could have been isolated every season in many parts of the world, though in some seasons they would have been so thinly distributed as to be locally negligible. Even in the leanest years the world total of influenza cases would have been found to be considerable had it been possible to identify them all. Moreover, in each epidemic the prevalent virus was attacking persons in all the age groups of the portion of the community that was nonimmune to it. Each epidemic was globally ubiquitous or very widely distributed, taking between six to twelve months to travel through the vast nonimmune portion of population, although in any particular locality it would have remained for only a few weeks or a month or two.

Each antigenic change, whether shift or drift, would have appeared contemporaneously in many communities living at a similar latitude, even when the epidemic was numerically small and the nonimmunes who were attacked were sparsely distributed.

In order to evaluate the task facing the epidemiologist attempting to explain the behavior of influenza A, one must also bear in mind the continuously changing age pattern of the human community in relation to the antigenic changes in the parasite. During the era of prevalence of a single major serotype, the persons immunized by its predecessors are aging and dying, and babies are being born and growing older. The average age expectancy of these babies in Great Britain will have spanned the eras of prevalence of several successive major serotypes. At any given moment the world population possesses a complicated pattern of influenzal immunity that is continuously altering season after season. Old people possess immunity against H2 and H3 subtypes and against all three major serotypes of H1 subtype and against influenza B virus, although in some of them the humoral antibody has waned.

The kaleidoscopic picture of population immunity to influenza A viruses is further complicated by the phenomenon of "original antigenic sin" whereby the current epidemic strain may elicit an even greater immune response to the first epidemic strain encountered in childhood. Thus the relatively simple pattern of immunity in childhood becomes progressively more difficult to interpret correctly as the young person grows older.

The diagrams in Figures 10.2 and 10.3 depict the behavior of each of the five major serotypes of human influenza A virus, and their immunological impact on the part of the world population nonimmune to each strain. The human community is depicted in 10-year age cohorts aging, dying, and being born during the 100 years previous to the writing of this book. The community from each successive influenza epidemic during an era of prevalence is denoted by the shading of the proportion of persons attacked and immunized in all the cohorts not already immune to that strain, namely, the persons who had been born since the end of the previous era of prevalence of that strain. By the end of each era almost all the nonimmune persons in each of the susceptible cohorts had been attacked and protected, many for life, from a further influenzal illness caused by that major serotype.

The five diagrams show closely similar behavior. Each major influenza A variant, whether major mutant or reassortant, appeared worldwide in its first epidemic season and in each subsequent epidemic. Each strain attacked persons in all age groups of the nonimmune community in each epidemic. Each major serotype virtually vanished at the end of the era of its prevalence, presumably because the world community had become too specifically immune to support further epidemics caused by that strain. All three subtypes returned, presumably when sufficient new nonimmune persons had been born.

The combination of the five diagrams into a single picture, were that possible, would illustrate the complexity of the pattern of immunity to influenza A viruses in the world population. Even so, the picture would be omitting the confounding effect of original antigenic sin, fortifying the immunity against the strain that had caused the first influenza A illness in each subject on encountering later strains of

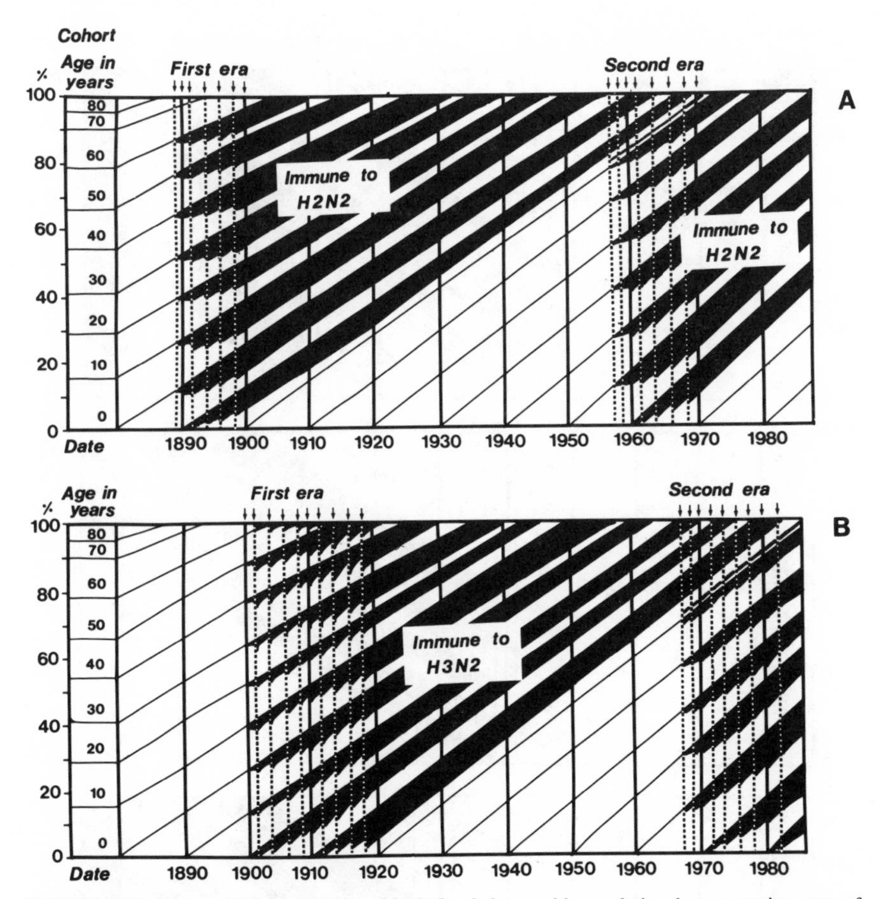

FIGURE 10.2. Sequential immunization (shaded) of the world population by successive eras of prevalence. (A) The A(H2N2) subtype. (B) The A(H3N2) subtype of influenza A virus. Note the long interpandemic intervals between successive eras of the same subtype.

influenza A virus. It would also be omitting the waning of immunity in some persons with advancing age.

FEATURES OF THE BEHAVIOR OF INFLUENZA A VIRUS THAT REQUIRE EXPLANATION

Although Figure 10.1 looks simple, it illustrates a number of features of the behavior of influenza A virus that have proved tantalizingly difficult to explain.

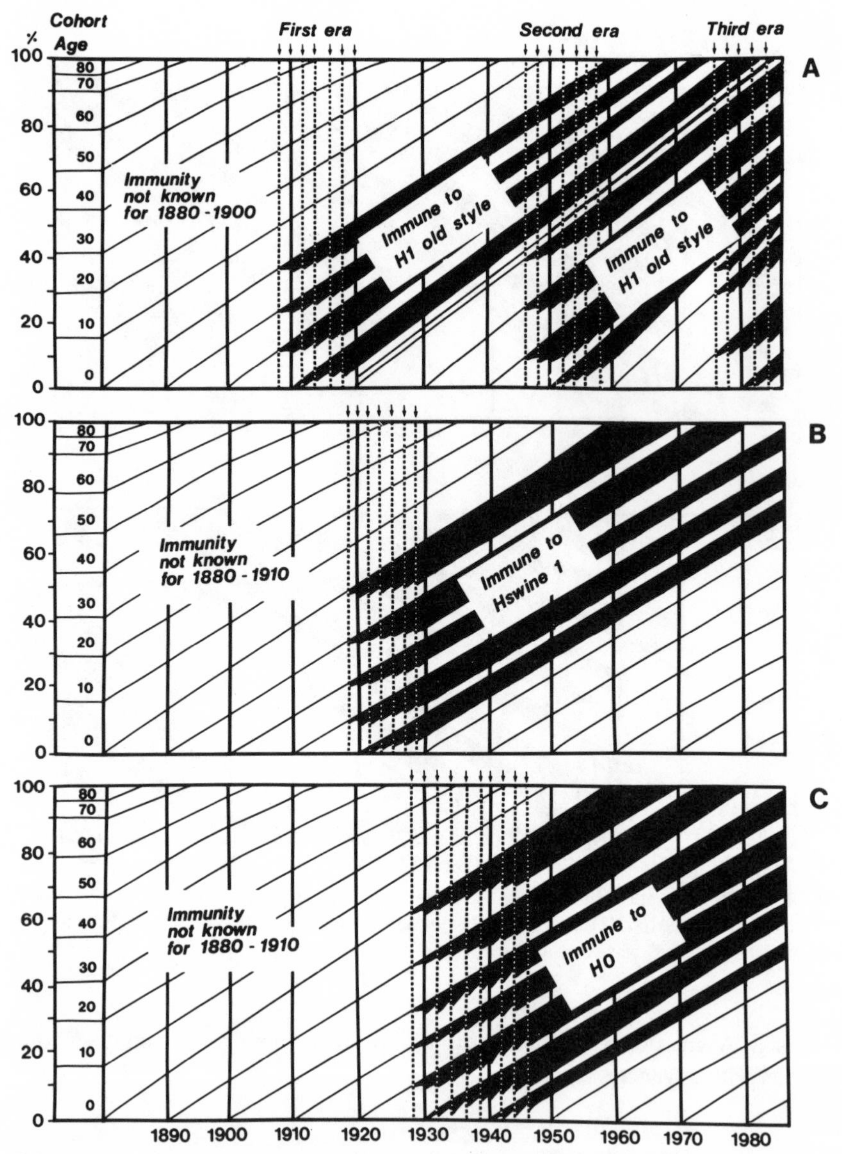

FIGURE 10.3. Sequential immunization by the A(H1N1) subtype is complicated by successive eras of prevalence of three major mutants during the first long prevalence of this subtype (see Fig. 10.1). (A) The A(H1N1 old style) variant which had three eras. (B) The A(Hswine1N1-like) variant. (C) The A(H0N1) variant.

They are listed separately, but some can only be usefully considered when others are also taken into consideration.

The Timing of the Major Mutations and Genetic Reassortments

In 1946 and 1957, it was assumed that the major antigenic changes were then taking place in the previously prevalent virus, somewhat in the manner of what is supposed to be happening at antigenic drift. Later observations make it unlikely that this is correct. It now seems probable that H1, H2, and H3 subtypes of human influenza A virus were produced by genetic reassortment many years or centuries ago and the same is probably true of the major mutants of H1 subtype known as H0, Hswine and H1 old style.

Recycling of Major Mutants and Subtypes

When Mulder and Masurel[2] discovered in 1956 that persons over 70 years old possessed antibody against A(H2N2) strains that were supposed to have appeared first in 1957, it seemed possible that the subtype had in fact had an earlier era of prevalence some half-century before. The discovery of a similar situation in 1968 when A(H3N2) strains were thought to have appeared as a novelty made it seem likely that both subtypes had been recycled in the order of their earlier prevalences. Then, when A(H1N1 old style) strains reappeared in 1977 after 20 years absence and when evidence came to light of an even earlier era of their prevalence in the first and second decade of this century, it seemed likely that recycling of major mutant serotypes and reassortment subtypes of influenza A virus must be occurring in mankind.

The findings are accepted by the new concept as evidence that these major variants were formed long ago, and must have been somehow stored for many years between their successive eras of prevalence in mankind. The most popular current hypothesis to explain their storage suggests that they are harbored in a nonhuman host species. The new concept proposes instead that they are stored in human hosts probably in the mode of the viral genome.

The Vanishing Trick

The vanishing trick was discussed in relation to antigenic drift in Chapter 9. Figure 10.1 shows that it is also characteristic of antigenic shift and of the major mutational changes now classified as drifts. In 1946, 1957, and 1968 the previously prevalent strain virtually disappeared from the world population as soon as the successor appeared.

It is not easy to propose an explanation. The solution proposed for the similar phenomenon at minor antigenic drift, namely, a sort of metamorphosis by which a novel variant is substituted for the previously prevalent strain, does not explain our present problem at these major antigenic variations of either sort because in such cases the novel strain has been recycled from storage and may be unrelated to its predecessor. A clue to the explanation of the vanishing trick may be found in the timing of these major changes at the moment when almost all persons in the world who were susceptible to the prevalent strain have become immune to it.

Rapid Replacement by the Successor

Here again one meets a problem already encountered at antigenic drift, though at antigenic shift it is on a larger scale. At each major antigenic change—1946, 1957, 1968, and 1977, and probably also 1900, 1908, and 1929—the successor appeared worldwide within a single season. The succession is too rapid to have been achieved by direct transmissions from the sick. Sources of the novel strain must already have been widely distributed globally before the major changes occurred.

The Location and Nature of the Stored Virus

For those who accept the serological evidence of the recycling of influenza A virus subtypes and major mutants, it is important to know where and in what form the virus is stored between successive eras of its prevalence. The length of time for which it is stored may suggest the mode in which the microorganism is retained, and the rapidity with which the recycled virus is distributed worldwide provides a hint as to the locations in which it has resided. As shown in Figures 10.1 and 10.2, H2 strains were stored for at least 57 years, H3 strains for 50 years, and H1 old style strains for 28 years and 21 years. Such lengthy absences followed by reappearance of the virus relatively unchanged suggests that the mode in which it has been stored was that of the viral genome. It is unlikely that the virus could have survived by a continuous chain of transmission for half a century without much evolutionary change, but it might well do so in the form of its RNA genome, thus retaining for an indefinite period the genetic code to reproduce the precise structure that it had before. The 1977 reappearance of A(H1N1 old style) strain illustrates this point. It was identical in all eight RNA segments with the Scandinavian strain, one of the two A(H1N1 old style) strains that had caused the great worldwide epidemic 25 years earlier.

If the duration of interpandemic absence suggests storage of the viral genome, the speed and ubiquity with which the resurrected virus appears throughout the world population excludes certain locations and suggests others. The "spread" is too rapid to have been initiated at a single location on the global surface. It has the seasonal character of other influenza epidemics occurring worldwide in a single season with about six months separating its first appearance in the two hemispheres. The explanation would seem to be the existence of ubiquitous available sources of the reactivating genome. Where can one find such sources ubiquitously available to the world population except in mankind? None of his common domestic animals lives long enough to harbor the genome from one of the eras of a human influenza A virus subtype until its next prevalence. It is true that the human strains that have infected swine have drifted antigenically much less rapidly than in their human host, but swine are not universally associated with mankind. The most probable site of genome storage must surely be the persons who have already suffered an infection by an influenza A virus of that subtype or major mutant.

The new concept therefore proposes that the location of the stored virus between successive prevalences is within the humans who were hosts to them in their earlier prevalence, and that the form in which it is stored is that of the viral RNA genome. The hypothesis further tentatively suggests that it may be only the first infection of the lifetime by an influenza A virus that results in such genome storage. This might explain the phenomenon of original antigenic sin. It would much reduce the number of potential sources of reactivating subtype genomes, but they would still be ubiquitously distributed in the world population.

Other hypotheses of antigenic shift will be discussed in the next chapter.

Solitary Prevalence of Major Influenza A Viruses

Another perplexing feature of the behavior of the major serotypes of human influenza A viruses has been the long periods during which strains belonging to a single major serotype almost alone have been causing all the influenza A in the world population. Figure 10.1 shows how H2N2 strains caused all the influenza A recorded from 1957 until the middle of 1968 and may have had a similar solo prevalence from around 1889 until 1900. Strains of H3N2 subtype caused almost all the influenza A from mid-1968 until 1977 and probably also from 1900 until 1908. H1N1 subtype strains had a similar solitary reign from 1918 until 1957.

These solitary prevalences have excited much comment. Types A and B influenza viruses readily co-circulated during all these periods and sometimes co-infected the same human host. No satisfactory explanation of the exclusive influenza A subtype dominances has yet been offered.

Subtypes are not unique in such exclusive behavior. Figure 10.1 shows that all three major mutants within the A(H1N1) subtype have had eras of solitary prevalence: H0 strains from 1929 until 1946, H1 old style strains from 1946 until 1957, and H-swine-like strains probably from 1918 until 1929. Indeed, for about 30 years after the first major antigenic variation was encountered in 1946 it was believed that some strange rule among influenza A viruses precluded the co-prevalence of major serotypes. The return of H1N1 strains in 1977 for an era of co-prevalence with the current H3N2 strains came as a surprise. Serological studies have discovered only one other such co-prevalence during the last 100 years, namely the similar co-prevalence of H1N1 and H3N2 strains around 1908–18. In both these co-prevalences it was the H1N1 old style major mutant of A(H1N1) subtype that participated with H3N2 strains.

A(H3N2) strains seem twice to have excluded A(H2N2) strains. The exclusions suggest that the subtypes and major mutants were developed long ago and have established a more or less ordered recycling within the human host species. Figure 10.1 indicates the possibility that A(H1N1) strains may have developed independently of the other two subtypes as a second lineage. The two lineages are distinguished by the possession of different neuraminidase-coding genes. No co-prevalence of major serotypes of A(H1N1) subtype has yet been recorded.

The Source of the Novel Genes at Antigenic Shift

At the antigenic shift in 1957, the A(H1N1) subtype strains that had been prevalent since 1908 appeared to have exchanged four of their eight RNA segments and so become the A(H2N2) subtype. In 1968, only one RNA segment had been exchanged, transforming A(H2N2) into A(H3N2) subtype. Reason has already been given to believe that these exchanges of genetic material cannot have been occurring in 1957 and 1968. The apparently novel H2N2 subtype must have been assembled long ago and stored and recycled perhaps countless times before the 1957 episode, and the same applies to the 1968 variation. Whence were the new H2- and H3-coding genes acquired originally and how were they incorporated into the genome of the human influenza A virus?

A vast reservoir of influenza A viruses has been found parasitic in many species of mammal and bird, both domestic and wild. Mankind may have originally obtained its influenzal parasites from this source and may still be receiving contributions. The discovery of the facility of genetic reassortment when different influenza A viruses co-infect a host focused attention on the mobility of individual genes between host species in addition to the mobility of intact viruses.

The finding of relationships between the surface proteins hemagglutinin and neuraminidase of some human influenza A subtype strains and those of the strains isolated from natural infections of certain animals and birds encouraged the belief that nonhuman hosts of influenza A virus might be contributing genetic material that participates in the epidemiology of human influenza A. The currently popular hypothesis of antigenic shift in human influenza A virus proposes that the novel genes are derived from avian and mammal strains of the virus.

The new concept accepts that genetic reassortants involving influenza A viruses of nonhuman host species may well have been the originals of the subtype, but proposes that the present behavior of influenza A in mankind is incompatible with dependence on regular or frequent involvement of the influenza viruses of nonhuman hosts. The antigenic shifts of 1957 and 1968 are proposed as having taken place entirely within the human community.

SEASONAL CHARACTER OF MAJOR ANTIGENIC CHANGES

The seasonal occurrence of antigenic shift is not evident in the illustrations. The antigenic shifts of 1957, 1968, and 1977 all occurred seasonally. Mention has been made of the seasonal change in antigenicity that characterizes the small mutations that so commonly cause antigenic drift, and the more major mutants in the A(H1N1) subtype also appeared seasonally. This similar seasonal pattern suggests that the same machinery that operates the timing of drift is at work eliciting shift.

If the recycling of subtypes and major serotypes depends on the recall of stored viruses—whether stored as genomes or in some other mode of latency—some mechanism must be postulated to operate the recall. The seasonal pattern that characterizes the reappearance both of vanished subtypes and of H1 major serotypes is the familiar seasonal pattern of all influenza epidemics. There must be a strong probability that the seasonally mediated stimulus that operates the recall of persistent virus in persons who have recovered from influenza to become highly infectious carriers (see Chapter 9: The Current Explanation of Antigenic Drift) is also operating the recall at antigenic shift.

UBIQUITY OF EPIDEMICS OF INFLUENZA

The ubiquity of influenza A has been mentioned several times, but it is such an extraordinary feature of viral behavior that it deserves more attention. The global distribution of the newly shifted virus subtype within a single season has been described and compared with the wide distribution of a new minor variant throughout the area of its predecessor at antigenic drift. But the ubiquity of epidemic influenza is more general. Influenza is an annually seasonal event worldwide. Almost every season strains belonging to a single influenza A subtype can be detected globally, although in some seasons the yield will be small. But even in these lean influenza years the strains will be widely distributed. Except in remote communities the strains isolated throughout the world will usually be up to date, as if each new major or minor variant was always able to distribute itself through the human population at the same speed as the solar annual cycle, but about six months after local midsummer. The behavior seems to be unique among parasites and may not be similar in nonhuman hosts of influenza A virus.

It is difficult to see how this ubiquitous seasonal epidemicity can be explained on a hypothesis of direct transmission from the sick.

THE TASK FACING THE EPIDEMIOLOGIST

The epidemiologist has the daunting task of identifying or suggesting the mechanisms whereby each subtype or major variant of influenza A virus first immunizes the whole of the previously nonimmune portion of the world population during the successive epidemics of its era of prevalence, then disappears from the world and is everywhere replaced next season by a novel major variant. The phenomenon always occurs at the opportune moment to save the human influenza A virus from becoming extinct because of lack of nonimmune subjects to support its continued survival. The orderliness of the process suggests that antigenic shift is an evolutionary adaptation that has been developed during centuries of close association between type A influenza virus and its human host in order that the parasite may evade the suicidal consequences of its immunogenicity and its high infectiousness. Another evolutionary adaptation that we shall be discussing may also have been developed to avoid these dangers, namely, the production of defective interfering particles that arrest the direct transmission of standard infectious virions from the patient sick with influenza, so delaying transmission until the human host has become immune and thus favoring antigenic drift, and at the same time limiting the spread of influenza to persons in contact with carriers in whom the virus is reactivating as described in the last chapter.

The epidemiologist is presented by influenzal antigenic shift with a cascade of related problems. How does antigenic shift come about? When and in what form is each such major variant abiding relatively unchanged during the years between the successive eras of its world prevalence? How is it summoned from its lair to renewed activity in mankind? What signal does it receive that the human community is again ripe for its return? When it returns, how does it become ubiquitous within a single season? Are the nonprevalent subtypes continually hammering on a closed door during their apparent absence?

Any epidemiological concept of the epidemic process of influenza A must simultaneously provide the answers to all these problems and to the numerous other puzzling features that characterize human parasitism by influenza A viruses if it is to be valid. The next chapter discusses some of the hypotheses that have been advanced in recent years.

The reader cannot have failed to be impressed by the number of phenomena that are common to shift and drift: seasonal antigenic change, the vanishing trick, rapid replacement. All the major and most minor variations have shown these features. One great difference is that drift is not usually repetitive. When the

A(H1N1) Scandinavian strain reappeared in 1977, the antigenic drifts from it were different from those of its 1951 predecessor.

Later we shall discuss anachronistic strains that reappear long after their prevalence. All reappearing strains, whether minor or major variants, must somehow have been stored in a situation in which neither mutation nor genetic reassortment was possible.

CYCLIC VARIATIONS IN SOLAR ACTIVITY AND INFLUENZA A VIRUS ANTIGENICITY

The discussion in Chapter 8 concluded that the rhythmic annual variations in solar radiation that determine all seasonal phenomena are therefore responsible for the seasonal epidemic behavior of influenza. We do not at present have any knowledge of the identity of the intermediate mechanisms whereby the annual variations in solar radiation exert their controlling influence on epidemic influenza, so that any further association of influenza with solar activity is a matter of particular interest.

The electromagnetic activity of the sun varies rhythmically in long irregular cycles, the activity increasing more rapidly than it declines. These solar cycles vary in length from 7 to 17 years, the average duration of a cycle being about 11 years.[3]

Sunspots on the solar surface have been recognized since the beginning of history and have been enumerated by astronomers for centuries. Their number gives an approximate guide to the state of the electromagnetic activity of the sun. Their numbers increase with the increasing electromagnetism and decrease in step with its decline. For reasons that need not be discussed here, the true cycle of solar activity is said to consist of two consecutive sunspot cycles measured from minimum to minimum.

A letter to *Nature* in 1978 drew attention to a remarkable concordance between sunspot maxima and the antigenic variation of influenza A virus. Six sunspot maxima had occurred between 1917 and 1971, and five of them had coincided closely with five successive major changes in the antigenicity of influenza A virus as shown in Figure 10.4 and Table 10.1.[4]

Subsequent experience may reveal that these striking concordances between a solar and a biological cycle are nothing more than an unusual coincidence, but until its random nature has been established it should be examined carefully as if it were a cause-and-effect phenomenon. The annual seasonal behavior of epidemic influenza and of the antigenicity of its viruses is undoubtedly ultimately attributable to the changes in solar radiation occasioned by the 23.5° tilt of the plane of rotation of the Earth in relation to the plane of its circumsolar orbit. This second

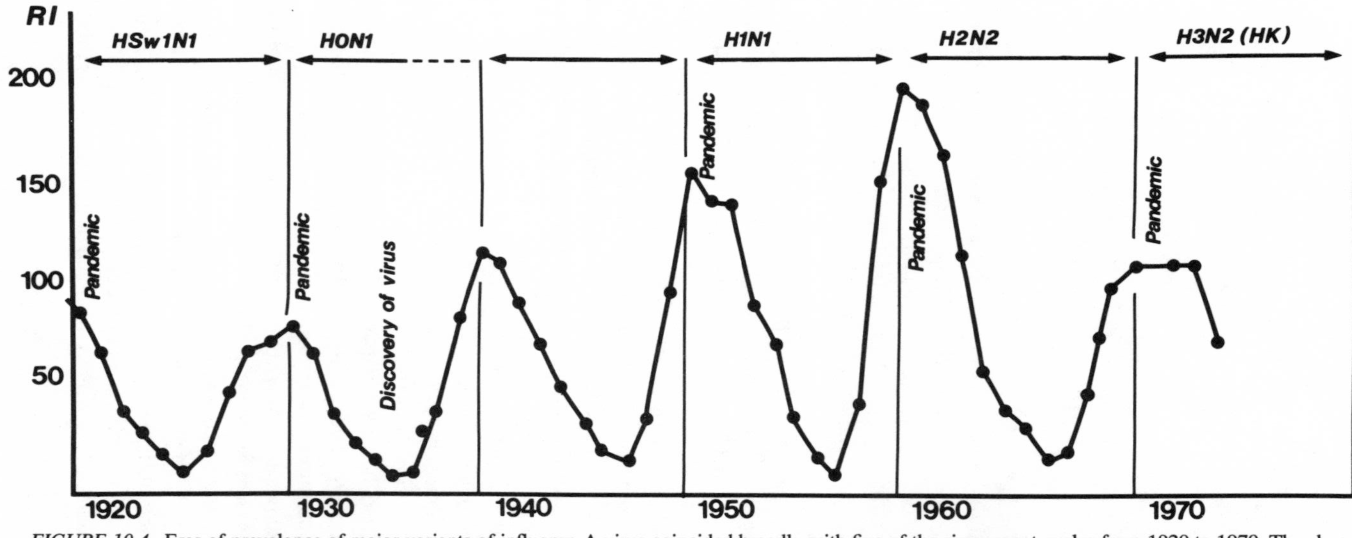

FIGURE 10.4. Eras of prevalence of major variants of influenza A virus coincided broadly with five of the six sunspot cycles from 1920 to 1970. The change of major variant occurred abruptly with the disappearance of its predecessor about the time of sunspot maximum. RI = Zurich yearly means of daily relative sunspot numbers (from Hope-Simpson[4]; reproduced with permission from *Nature*).

TABLE 10.1. Concordance between Sunspot Maxima and Date of Appearance of Major Variants of Influenza A Virus

Year	Major variant		Sunspot maximum
1908	H1N1 old style	now H1N1 subtype	
1918	Hswine1N1	now H1N1 subtype	1917
1929	H0N1	now H1N1 subtype	1928
			1936
1946	H1N1 old style		1947
1957	H2N2 subtype		1957
1968	H3N2 subtype		1968

relationship of influenzal behavior to an intrinsic solar rhythm affecting the composition and intensity of the solar radiation received by the Earth may be a source of much needed information on the mechanisms by which such biological effects are operating.

At the time that the letter to *Nature* was conceived, all five major antigenic variants were classified as antigenic shifts initiating changes of subtype. Now, however, it is certain that the first three variants in 1918, 1929, and 1946 were mutants of the subtype now called A(H1N1) influenza virus. The variants in 1957 and 1968 were both reassortments initiating the era of prevalence of strains belonging to a new subtype. Minor and major antigenic drifts both occur seasonally as do antigenic shifts, a fact that suggested that the variations in solar radiation were operating by similar intermediate mechanisms in both sorts of variation. The finding that a different solar rhythm of electromagnetic activity, a different source of variation in solar radiation of the Earth, may also be causing changes in the antigenic nature of influenza A viruses prevalent in the human population, whether mutants or reassortants, strengthens the case that a single chain of intermediate mechanisms is in operation.

The search must be for a single operation that would be common to antigenic drift and antigenic shift. The most likely effect to link these variations is a mechanisms for recalling viruses stored in some form of persistence or latency to renewed activity and infectiousness. It seems highly unlikely that these major variants have been formed at the date of their appearance. They are almost certainly reappearances of strains that were developed earlier, possibly centuries earlier, and they have somehow been reclaimed from storage.

The major variations have occurred in years when the Earth receives its maximum electromagnetic radiation from the sun, whereas minor antigenic variations may occur at the time of year when the solar radiation is lowest.

Hoyle and Wickramasinghe,[5] whose theory of influenzal transmission is discussed in Chapter 11, have recently examined the concordance of severe pandemic influenza with sunspot maxima since 1761. They also extend the illustration reproduced here as Figure 10.4 to include subsequent years up to 1990. They comment as follows:

> Although the correspondences between the figure and our table are clearly not precise, they do add credence to the speculation that solar activity may have a causal link. The two phenomena, which have irregular periods (with an average period of 11 years), appear to have kept in step over some 17 cycles.
>
> Past experience has shown that false correlation of phenomena with the sunspot cycle may look good over a few cycles, but go seriously adrift after an appreciable number of cycles. This does not happen for the postulated sunspot–flu connection.

They note that electrical fields associated with intense solar winds could rapidly drive charged particles of the dimension of influenza virions through the upper atmosphere (where they would be exposed to lethal radiation) to the lower atmosphere where they would be more sheltered. The photoelectric effects would have charged the virus particles. They see this as a possible link between influenza pandemic and solar activity that would be in keeping with their theory of influenzal transmission as outlined in the next chapter (see Table 10.2).

A word of caution is needed. As emphasized earlier, epidemic influenza is an annual phenomenon on the global scale, and it would therefore require to be

TABLE 10.2. Concordance of Sunspot Maxima with Serious Influenza Pandemics 1761–1893 According to Hoyle and Wickramasinghe[a]

Sunspot maximum	Date of pandemic
1761	1761–62
1767	1767
1778	1775–76
	1781–82
1787	1788–89
1804	1800–02
1830	1830–33
1837	1836–37
1848	1847–48
	1850–51
1860	1857–58
1870	1873–75
1893	1889–90

[a]From Hoyle and Wickramasinghe[5]; reproduced with permission from *Nature*.

shown that the annual curves of influenzal epidemic severity coincide approximately with those of solar activity. This may well be the case, but has yet to be demonstrated.

Van Alvesleben[6] has challenged the accuracy of the figures quoted by Hoyle and Wickramasinghe. Using the sunspot numbers published by Waldmeier, he shows that the sunspot maxima have been "fairly well distributed over all phases of the solar cycle." He also calls attention to the dubiety attending the distinction between epidemics and pandemics drawn from the older records of influenza.

REFERENCES

1. WHO memorandum. A revision of the system of nomenclature for influenza viruses. *Bull WHO* 58:585–591, 1980.
2. Mulder J, Masurel N: Pre-epidemic antibody against 1957 strain of Asiatic influenza. *Lancet* 1:810–814, 1958.
3. Gibson EG: Description of solar structure and processes. *Rev Geophys Space Phys* 10:395–461, 1972.
4. Hope-Simpson RE: Sunspots and flu: A correlation. *Nature* 275:86, 1978.
5. Hoyle F, Wickramasinghe NC: Sunspots and influenza. *Nature* 343:304, 1990.
6. von Alvesleben A: Influenza according to Hoyle. *Nature* 344:274, 1990.

11

Hypotheses of Antigenic Shift

HYPOTHESES OF SHIFT INVOLVING NONHUMAN HOSTS

In the wake of the great antigenic shift of 1957, which replaced A(H1N1) strains by A(H2N2) strains and initiated the Asian influenza pandemic, Fred Davenport[1] in the United States and Mulder and Masurel[2] in the Netherlands and later Andrewes[3] at Hampstead suggested that the novel virus had been derived from an influenzal infection of some southeast Asian mammal or bird. Not much was then known about influenza in nonhuman hosts and the speculations are reminiscent of those made by Fothergill and others in the eighteenth century who suspected that human influenza epidemics might be extensions of epizootics in domestic animals such as dogs and horses.

Davenport and the others did not have to wait long for information about influenza in birds and nonhuman mammals, and much of it seemed to support their speculations. Influenza virus infection of nonhuman hosts, far from being rather rare, was found to be so common that there is a vast reservoir of influenza A viruses in wild and domestic birds and mammals.

The zoonotic origin of human influenza A virus subtypes gained credibility when similarities were found between the hemagglutinins and neuraminidases of some strains of influenza A viruses of mankind and those of birds and mammals. Many more such interspecies antigenic relationships seemed likely to come to light as knowledge about the animal influenzas was extended.

Some of the findings suggested that antigenic shifts of human influenza A virus might have arisen by reassortment between the H-coding (and sometimes also the N-coding) gene of human and animal strains rather than by transfer of whole virions from animal to human host. A number of workers including Kilbourne[4] and Laver and Webster[5] showed that such genetic reassortments could be produced experimentally. If a domestic turkey or chicken was simultaneously infected in different parts of the respiratory tract with two different strains of influenza A virus, both of the original strains and one or more hybrid viruses could be isolated during the resultant illness of the bird.

The concept of a natural animal reservoir for generation of human influenza A subtypes gained added credence when it was found that the Asian A(H2N2) strain that appeared in 1957 had retained only four of the eight nucleoprotein segments of its A(H1N1 old style) predecessor and had acquired from some unidentified source four novel genes including those coding for hemagglutinin and neuraminidase. The finding generated intense speculation as to the origin of the novel genes and the process whereby they had been incorporated into the predecessor's chromosome. The novelty of the surface antigens, hemagglutinin and neuraminidase, to which only some aged persons possessed any immunity, explains the catastrophic impact on mankind of the antigenic shift of 1957.

The next antigenic shift in 1968 involved only the change of the gene coding for hemagglutinin, the other seven segments of the A(H3N2) strains being conserved from the A(H2N2) predecessor. The H3 antigen was found to be similar to those in both an equine and an avian strain.

KILBOURNE'S 1975 HYPOTHESIS

These similarities between human and nonhuman influenza A viruses may be indicating spread from nonhuman to human host, spread in the opposite direction, or spread in both directions. In 1975, Kilbourne[6] expressed his belief that there are no true animal strains of influenza virus. At that time the influenza viruses of other than human hosts were named from the host species from which they had first been isolated as, for example, A/equine, A/swine, and so on. Kilbourne considered that all animal hosts had initially derived their influenzal parasites from mankind. The pig, for example, domesticated for 9000 years, had had ample time to have evolved a relationship with its master's parasites.

Humans must have been transmitting their strains of influenza virus now and then to their most closely associated domestic creatures. More rarely such transmissions would have resulted in adaptation of the human strain to continued parasitism of the new host species, as in 1918 when domestic pigs in the United States are thought to have retained the A(Hswine1N1-like) strain caught from human farm workers suffering from the so-called "Spanish 'flu." Similarly, since 1968, human A(H3N2) strains have become enzootic and sporadically epizootic in swine in many parts of the world,[7] and the A(H1N1 old style) strains that returned to a new era of human world prevalence in 1977 also seem to be becoming established as a parasite of the domestic pig.[8]

Kilbourne surmised that adaptation to an alternative host unfitted the human strain for parasitism in mankind so that it seldom makes the return journey. He went on to speculate that, as a very rare event indeed, human and animal viruses, meeting either in man or in another host, may exchange genes and so produce a hybrid capable of bypassing human immunity and causing a pandemic "if an ecological niche is available." He notes that Nature has endless time for her experiments, and he has the alarming vision of these pandemic candidate influenza viruses already in existence lining up for the opportunity to gain or regain access to the human population.

Kilbourne emphasized two important inferences to be drawn from the then available findings: (1) major subtype variants can be derived from antecedent human subtypes, and (2) antigenic variations in the hemagglutinin are limited in number, meaning presumably the major variations and perhaps referring only to human strains.

He admitted that it seemed unlikely that the two appearances of A(H3N2) strains in 1900 and 1968, in each of which the hemagglutinin resembled that of an equine strain, could both have been derived from an equine host. He also proposed, what is now generally accepted, that the major H1 antigenic variants of human influenza A virus from 1918 to 1947 were mutants whereas the large variations in 1957 and 1968 resulted from genetic recombination.

Kilbourne's fascinating discussion was published in 1975 just before the events of 1976 and 1977 added so much to the available knowledge that they altered the views of many people. He had considered that the two eras of A(H2N2) strains—1889 to 1900 and 1957 to 1968—and of A(H3N2) strains—1900 to 1918 and 1968 to the present day—were an insufficient periodicity to support a concept based on the recycling of subtypes of influenza A virus. The concept of ordered recycling was, however, much strengthened when, two years after Kilbourne's textbook was published in 1975, A(H1N1 old style) strains returned in 1977 for a renewed era of prevalence, which Masurel and Heijtink[9] have shown serologically to have been a third era of that variant during this century.

When discussing the problem of the occurrence of pandemics and of the vanishing of the preceding subtype, Kilbourne pointed out that these are phenomena peculiar to influenza A virus. Other viruses and bacteria manage to survive in highly immune populations without the need of major antigenic changes—influenza B virus for example. He concluded that influenza A virus must be walking a precarious tightrope in sustaining parasitism of man, only able to balance if a high proportion of nonimmune subjects is available in the community.

Kilbourne's conclusion is based on the assumption that the virus must survive by direct spread. The need for a high proportion of nonimmunes is not supported by our own experience. Unlike some other epidemic parasites, influenza A virus can mount a major epidemic in a largely immune community, as in the case of A(H2N2) strains in their final epidemic in the Cirencester community during the first four months of 1968. In this eighth considerable outbreak since their arrival in 1957, they caused the second largest of the epidemics despite the high proportion of immune persons in the local population, and they attacked people of all ages.

The findings are among a number that indicate that influenza epidemics should not be modeled mathematically by the methods developed for the study of measles. There seems to be no evidence of the need for a critical proportion of nonimmunes in the community before an epidemic of influenza can erupt. The similarity to chemical mass action noted in measles does not seem to apply to influenza epidemics during an era of subtype prevalence.

Although small and large influenza epidemics occur independently of the proportion of nonimmunes until almost the whole community has been immunized, the initiation of a new era of prevalence is probably much influenced by the proportion of nonimmunes. The subtype eras of influenza in this way resemble individual epidemics of measles.

Kilbourne,[10] in 1973, had proposed an ingenious hypothesis of how the domestic pig might be acting as a reservoir where human influenza A virus subtypes are being preserved with little or no antigenic change, and how the pig could thus also provide the location where animal and human influenza A viruses from human and nonhuman hosts could meet and generate novel subtypes by reassortment and recombination of their genes. The new strains thus produced could then find their way into circulation in mankind as in Figure 11.1.

I am indebted to Kilbourne for permission to reproduce his diagram of how the domestic pig might act as a reservoir in which human subtypes of influenza A virus could be preserved with little antigenic change, and of how it could also provide a location where genetic recombination and the generation of novel subtypes could occur.

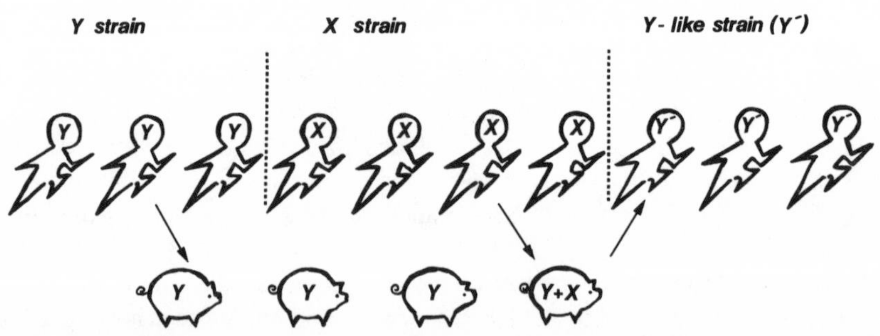

FIGURE 11.1. Kilbourne's hypothesis of shift: human strain Y rarely infects, say, the domestic pig and becomes established as a porcine strain. At the end of its era of human prevalence, Y disappears from mankind and is succeeded by strain X. Even more rarely, human strain X infects a pig already infected with porcine strain Y. Recombination between X and Y may produce a strain antigenically resembling Y but with X's ability to replicate in man—call it Y′. Y′ may then appear as a "new human influenza virus." Hundreds or thousands of years may separate such events. The new concept accepts this as one possible mechanism whereby the human subtypes originated (from Kilbourne[26]; reproduced with permission from Natural History Publications).

THE HYPOTHESIS OF WEBSTER AND LAVER IN 1975

Robert G. Webster and W. Graeme Laver[11] have summarized the evidence that the antigenic shifts of 1957 and 1968 were caused by recombinants and not by mutations in the predecessor strains, but when did these recombinations occur? Did they happen in 1957 and 1968, or did the recombinations take place years or even centuries before?

Although serological archaeology had already suggested that subtypes had been recycled, Webster and Laver noted that the N2 neuraminidase antigen of the 1968 A(H3N2) strains was not precisely identical with that of the earlier A(H3N2) strains prevalent from 1900 to 1918. The small difference in the N2 antigens caused them to doubt the evidence for recycling, or at least the hypothesis that the virus or its genome had been retained within the human host species between the successive eras of prevalence of each subtype.

In their view recycling might have resulted from the reemergence of the previous strain from a reservoir in some nonhuman host, either with or without genetic recombination, when human herd immunity no longer precluded it. Their reason for suggesting that the virus is being stored in a nonhuman animal reservoir rather than in a human host seems to have been the similarities that they and others had found between the H-and N-coding genes in human and animal strains.

They attributed the vanishing trick at antigenic drift—the disappearance of the previous strain—to "self-annihilation" because original antigenic sin would have enhanced earlier antibodies and so prevented spread of earlier strains. The ingenious explanation is an example of the unacceptable use of the hypothesis of herd immunity, because it is chiefly nonimmune persons who are attacked by the novel minor variants at antigenic drift.

In discussing the similar vanishing trick at antigenic shift they admit candidly that "no satisfactory explanation is available." When such an unusual and apparently illogical phenomenon affects two closely related processes, it would seem wise to seek an explanation common to both. In the last chapter we discussed this difficult problem.

In 1977 Fred Davenport[12] expressed skepticism about the existence of reservoirs of human strains in nonhuman hosts as explaining the behavior of influenza A in mankind. He pictured a more or less orderly recycling as reflected in the pattern of influenza A virus antibody in the human community. He, however, admitted that there are numerous problems still awaiting further information.

LATER CONCEPTS OF KILBOURNE

Kilbourne[13] published his matured views in 1987 in a fine book simply entitled *Influenza,* the harvest of a lifetime of work and cogitation. He has "attempted to make some sense of this baffling disease and its viruses," and to reexamine old questions in the light of "the dazzling advances in molecular biology." He remarks that, although no virus has been better studied, few diseases are less well understood. As a physician who has made outstanding contributions to the subject he is galled by this ignorance. His book is based on much personal experience, and wide and deep knowledge of the literature of influenza.

His view of the continuing evolution of influenza virus is stimulating. He sees the separate identities of the viruses as, in the long view, ephemeral. The influenza viruses are transiently stable packages of genes borrowed from an extended gene pool. We should be studying not the viruses but the evolution of the genes. Each gene is destined to evolve, not in a single package (i.e., a strain of the virus), but moving by recombinations from package to package. A mutation occurring in a gene may therefore be tested in numerous genotypes so that it will be selected "according to its average fitness rather than according to the fitness of the genotype in which it originated." Kilbourne considers that by this device influenza viral genes obtain benefits that some other creatures obtain by the device of sexual reproduction. The segmented genome of rather loosely associated genes of influenza viruses allows them to share their individual divergent evolution in the extended pool of influenza genes existing in numerous different animal hosts in various ecological niches. He recognizes this deployment of genes in varied hosts and circumstances as the evolutionary strategy of influenza virus. Wherever they happen to be residing, the genes can be recruited for parasitism elsewhere as population immunity or chance interspecific host contacts provide opportunity.

Evolutionary development into types A, B, and C has limited "this virological ecumenism, and even among genetically homologous influenza A subtypes, gene sampling is probably infrequent."[13] Thus Kilbourne explains both the plasticity and the relative stability of influenza viruses.

Kilbourne mentions the new concept advanced in this book but dismisses it as unnecessary because he maintains that the virus is able to survive between epidemics by continuous sequential transmission from person to person at a level below the "epidemic threshold," a conception borrowed from the measles model, many of the infections being symptomless. Yet he fails to offer any explanation of the rhythmic seasonal changes from the few weeks of acute epidemicity to the many months of his proposed low-level endemicity and back again to epidemicity. He also fails to explain why epidemicity and its accompaniment of changes in antigenicity occur seasonally.

He also rejects the new concept proposed here because he claims that there is no evidence that influenza viruses can adopt any mode other than that of acute infectiousness. It is difficult to see how he can sustain this argument. Latent infection by A(H1N1) and A(H3N2) strains has been shown by Gourreau and his colleagues[14] to occur naturally in swine epizootics, and Russian workers have produced it in laboratory mice.[15] We shall be discussing the abundant evidence of persistent infection and latency of influenza A virus in cell cultures. Jakab and his colleagues[16] found antigens of the virus in pulmonary alveolar cells of infected mice months after they had recovered from influenzal pneumonia.

Kilbourne provides a clear account of the progressive changes in the antibody status of the community, which he terms the "population phenotype." When considering this human herd immunity he omits to point out that it is mainly nonimmune persons who are attacked by the prevalent virus, and that they comprise *the only population group that cannot be exerting the proposed immune pressure for antigenic change* attributed to herd immunity. He is not alone among theorists in this important omission.

How does the influenza virus survive during the months of its apparent absence between epidemics? Surveillances have isolated the virus in every calendar month and have noted antibody rises in symptomless persons that have been attributed to asymptomatic infections. Kilbourne considered that these occasional isolations and symptomless infections, if that is what they are, are adequate to explain virus survival from one epidemic to the next, although he admitted that there is no evidence that the virus can be transmitted form a subclinical infection. It is unfortunate that he leaves the problem there, as if everything had been explained by some sort of low-level chain of transmissions. Champions of the current concept of direct spread must also explain how and why pathogenicity should rapidly dwindle to negligible proportions after a few epidemic weeks, how and why it should remain imperceptible for many consecutive months, and how and why the virus can then abruptly regain high pathogenicity and epidemicity in the same community. If direct spread is occurring, epidemicity must be dwindling and increasing *pari passu* with pathogenicity, because the nonimmunes in the population would be used up in a single season unless the number of persons infected between epidemics fell far below the epidemic incidence. One can postulate mechanisms to bring about such a reduction. For example the serial interval might be much protracted between causally related interepidemic infections. But this would leave much to be explained about the nuts and bolts of the process and, in fact, no attempt seems to have been made to provide an explanation of this aspect of influenzal epidemic behavior, nor of why it is broadly characterized by seasonal timing and is therefore determined by the latitude at which the community is situated.

In 1981, Monto and Maassab[17] published serological evidence that "vanished" subtypes of influenza A virus continue to circulate in the human community at low prevalence, and that their reappearance for an era of prevalence is signaled by an increase of antibody to them in the community shortly before they emerge from subdued prevalence. Their findings lend support to the anthroponotic concept because, if the subtype strains are continuously present, albeit scanty, there is no need to postulate storage in an alternative host species between eras of prevalence. We still need to explain why and how an intensely infectious parasite abruptly switches to subepidemic endemicity, perhaps for 50 years, then equally abruptly reverts to epidemicity. Endemicity by continuous direct spread does not solve the difficulties and raises new problems demanding explanation.

Despite his skepticism, Kilbourne advises that the possibility of influenza virus adopting some mode of latency ought not to be dismissed,[18] and it therefore seems reasonable to include it tentatively as one of the features of the new concept that can explain so many otherwise inexplicable aspects of human epidemic influenza.

THE SIGNIFICANCE OF ANTIGENIC DIFFERENCE OF NEURAMINIDASE IN OLD AND NEW STRAINS

There has been much speculation concerning the place in which the influenza A virus or its genome is stored interpandemically, for example, during the half-century of absence of A(H2N2) strains from 1900 until 1957 and the hiding place of the other subtypes during their long absences. In what form are they preserved and how are they recalled?

Some workers consider such questioning unnecessary. They regard each new era as having been initiated by a new genetic reassortment involving at least the gene coding for the hemagglutinin.

As mentioned earlier, the neuraminidase of the 1968 A(H3N2) strains differed a little from the N2 of the 1900 A(H3N2) strains, a finding that seemed to support the hypothesis that the two A(H3N2) strains had had separate origins.

But finding that the neuraminidases are distinguishable does not necessarily carry such a significance. The evidence about the earlier strains came from the serology of elderly persons in many of whom antigenic drift during the era of prevalence must have altered the genome by minor mutations to a degree that varied according to the year they were infected between 1900 and 1918. The hemagglutinins seems to have been identical in the two eras.

A useful lesson may be learned from the two prevalences of A(H1N1 old style) strains that have occurred since the discovery of the virus in 1933. Both prevalences could be studied virologically as well as serologically. The novel 1977 strains were distinguishable from the earliest 1947 strains and from the later strains of that earlier era after 1953, but they were identical in all eight RNA segments with the Scandinavian strains of 1950–51 season.[19] Soon after the 1977 strains appeared they began drifting antigenically in a direction entirely different from that followed by their 1951 forerunner.

If the strain of the previous era of prevalence has been maintained within the human host throughout the interpandemic period, there should be no surprise that retrospective serology of persons affected by that strain in the previrological era finds the earlier neuraminidase antibody differing in a minor degree from that of strains isolated in the later era. The finding cannot be taken as evidence that recycling has not occurred.

HYPOTHESIS OF AN INFLUENZA EPICENTER IN CHINA

Several influenza epidemics are thought to have originated in China. Dr. Norman White, who was for many years epidemiological adviser to the government of India, was commissioned in 1919 to undertake a study of the impact of the 1918 influenza pandemic, not limiting his report to the subcontinent of India. For reasons that are now obscure, he and others considered that in order to avoid a recurrence of such pandemics it was necessary to undertake the dredging of the Yangtse river and the Yellow river in China.

At that time White was unaware that influenza viruses existed and were parasitic in nonhuman hosts. Epidemiologists have recently offered the more plausible suggestion that somewhere in China south of the Yangtse river an "influenza epicenter" must exist where man, pig, and domestic ducks live in close association.[20] Ducks are known to be able to harbor all three human influenza A virus subtypes in addition to any of the 27 other known combinations of hemagglutinin and neuraminidase antigens. The influenza A viruses inhabit the gut of healthy ducks, which automatically distribute a motley assortment of influenza A virions in their droppings into the water and mud in the paddy fields, and in and around the numerous leets and duckponds. They may thus share their influenzal parasites with any available creature that can provide suitable host conditions, and what host would be more likely to be a recipient than the domestic pig that rootles in the soil, wallows in the mud, eats the earthworms, and drinks the contaminated pondwater?[21]

The proposed epicenter could be providing an ideal situation where human strains transmitted to pigs could encounter other influenza viruses that happened to be present in the pig including the avian strains received from ducks in the manner described. The viral mélange in the pig might allow swine and human strains to be transmitted from pig to duck and back again to the pig. Recombination between the influenza virus chromosome of the different strains in either or both domestic animals might provide novel influenza A viruses that could occasionally be transmitted from the pigs to their human companions, usually without effect either because the recombinant could not adapt to human parasitism or because that ecological niche was already occupied.[22] Nature, however, is patient. Sooner or later, so the hypothesis suggests, and adaptable strain will be transmitted when the immune situation in the human community is favorable to the novel recombinant, and a pandemic will presage an era of its worldwide human prevalence.

The attractive hypothesis provides a location for the origin of new subtypes of human influenza A virus where the mechanism of their production in animal hosts and for their transmission to man as suggested by Kilbourne might well be taking place.

There are, however, serious difficulties. How does the novel strain acquired by a man from a pig in rural China achieve world distribution within a single season? It could not do so by means of direct spread from cases of human influenza.

The new concept explains the origin of the three subtypes of human influenza A virus by mechanisms similar to that proposed for the Chinese epicenter, but considers that these must have happened at various times many years or even centuries ago. Their initial distributions throughout the world population were probably slow, occupying many influenza seasons, but, when each had been thoroughly distributed, their successive eras of human prevalence seem to be best explained by storage of the genome within persons who had been infected. The only species of domestic animal universally distributed in close contact with man is his human companions. Some mechanism has to be invoked for reactivating the genome in such human carriers.

Perhaps the pandemics that began in the Far East during the nineteenth and twentieth centuries did so because that is where the subtypes originated centuries ago and the pattern was set for all time. The more or less orderly recycling may also be reflecting the order and timing of the production of subtypes and other major serotypes. They possess idiosyncrasies that, as the years pass, bring them out of step with one another, because durations of eras of prevalence and interpandemic absence differ from one serotype to another. The interplay between viral antigenicity and human herd immunity must be important to the ordered recycling of eras of major serotype prevalence.

A DUAL RECOMBINATION HYPOTHESIS

In 1979, Dr. Wang Mau-Liang[23] of Wuhan University in the People's Republic of China suggested that human influenza A epidemics are initiated by a double recombination event affecting influenza A virus. First, each subtype may be capable of existing within the cells of the infected human host in the mode of its noninfectious genome, a suggestion that accords with a proposal of the new concept. After many years, conditions may offer the opportunity for the latent genome to recombine with another human or animal influenza virus, so creating a novel pathogenic strain. Some such strain might be available to cause a pandemic at a time when the immune status of the community was favorable.

The proposal that the genome can be retained for indefinite periods within the cells of the carrier seems sufficient to account for the antigenic shifts that have occurred. Dr. Mau-Liang, however, may be proposing the "rescue" mechanism whereby the latent genome is revivified by encounter with an infective virion. This valuable suggestion is not quite clear from his letter to the *Lancet.*

REFERENCES

1. Davenport F: The clinical epidemiology of Asian influenza. *Ann Intern Med* 49:493–501, 1958.
2. Mulder J, Masurel N: Pre-epidemic antibody against 1957 strain of Asiatic influenza. *Lancet* 1:810–814, 1958.
3. Andrewes CH: The epidemiology of influenza. *J R Soc Health* 78:533–536, 1958.
4. Kilbourne ED: Recent contribution of molecular biology to the clinical virology of Myxoviruses. *Yale J Biol Med* 53:41–45, 1980.
5. Laver WG, Webster RG: Studies on the origin of pandemic influenza. III Evidence implicating duck and equine influenza virus as possible progenitors of the Hong Kong strain of human influenza. *Virology* 51:383–391, 1973.
6. Kilbourne ED: Epidemiology of influenza, in Kilbourne ED (ed): *The Influenza Viruses and Influenza.* New York, Academic Press, 1975, pp 483–538. See especially pp 513–517.
7. Ottis K, Sidoli L, Bachmann PA *et al:* Human influenza A virus in pigs: Isolation of a H3N2 strain antigenically related to A/England/42/72 and evidence for continuous circulation of human viruses in the pig population. *Arch Virol* 73:103–108, 1982.
8. Masurel N, de Boer GF, Anker WJJ *et al:* Prevalence of influenza viruses A-H1N1 and A-H3N2 in swine in the Netherlands. *Comp Immunol Microbiol Infect Dis* 6:141–149, 1983.
9. Masurel N, Heijtink RA: Recycling of H1N1 influenza A virus in man—a haemagglutinin antibody study. *J Hyg* (Camb.) 90:397–402, 1983.
10. Kilbourne ED: The molecular epidemiology of influenza. *J Infect Dis* 127:478–487, 1973.
11. Webster RG, Laver WG: The antigenic variation of influenza viruses, in Kilbourne ED (ed): *The Influenza Viruses and Influenza.* New York, Academic Press, 1975, pp 270–314.
12. Davenport F: Reflections on the epidemiology of Myxovirus infections. *Med Microbiol Immunol* 164:69–76, 1977.
13. Kilbourne ED: *Influenza.* New York, Plenum Medical, 1987.
14. Gourreau JM, Kaiser C, Madec F *et al:* Passage du virus grippale par la voie transplacentale chez le porc dans les conditions naturelles. *Ann Inst Pasteur* (Virol) 136E: 55–63, 1985.
15. Zuev VA, Mirchink EP, Kharitonova AM: Experimental slow infection in mice. *Vopr Virusol* 24:29, 1983.
16. Jakab GJ, Astry CL, Warr GA: Alveolitis induced by influenza virus. *Am Rev Respir Dis* 128:730–738, 1983.
17. Monto AS, Maassab HF: Serologic responses to nonprevalent influenza A viruses during intercyclic periods. *Am J Epidemiol* 113:236–244, 1981.
18. Kilbourne ED: Molecular epidemiology—Influenza as the archetype. *Harvey Lect* 73:225–258, 1978.
19. Kendal AP, Noble GR, Skehel JJ *et al:* Antigenic similarity of influenza A(H1N1) viruses from epidemics in 1977–1978 to "Scandinavian" strains isolated in epidemics of 1950–1951. *Virology* 89:632–636, 1978.
20. Shortridge KF, Stuart-Harris CH: An influenza epicentre? *Lancet* 2:812–813, 1982.
21. Hinshaw VS, Webster RG, Turner B: Novel influenza A viruses isolated from Canadian feral ducks; including strains related to swine influenza. *J Gen Virol* 41:115–127, 1978.
22. Hinshaw VS, Bean WJ, Webster RG *et al:* Genetic reassortment of influenza A viruses in the intestinal tract of ducts. *Virology* 102:412–419, 1980.
23. Mau-Liang W: Dual recombination as origin of pandemic influenza viruses. *Lancet* 2:1077, 1979.
24. Kilbourne ED: Man's adaptable predator. *Nat Hist* 82(1):72–77, 1973.

12

Some Other Epidemiological Hypotheses

THE THEORY OF A BLAST FROM THE STARS

In Chapter 1 it was mentioned that two scientists had recently proposed a theory suggesting that the simile of Thomas Willis that the 1658 influenza epidemic had come like a blast from the stars may have been nearer the truth than he had realized. Sir Fred Hoyle, astrophysicist, and Professor Chandra Wickramasinghe, mathematician, working at University College, Cardiff, South Wales, found themselves unable to explain the behavior of epidemic influenza in the local day schools by direct spread of the virus from the sick. Earlier studies had convinced them that terrestrial life had originated not, as generally supposed, on the Earth itself but in a sort of premetabolic soup in the tails of the abundant comets that are roaming the galaxy. They consider that pristine microorganisms attached to cosmic dust particles are continually being scattered at random into space from such sources, and those that happen to be distributed appropriately are being wafted onto the Earth in the electromagnetic stream known as the solar wind. It is outside the scope of this book to discuss the merit of their theory of the extraterrestrial origin of life on the Earth, but the corollary of the theory as they have applied it to epidemic influenza must be described and discussed.

As a result of their inability to explain influenzal behavior by the current concept of direct spread, Hoyle, Wickramasinghe, and more recently Dr. John Watkins, another physicist now a general medical practitioner, extended their cometary theory to explain not only the behavior of influenza but also that of measles, mumps, whooping cough, and smallpox. Not unnaturally their hypotheses have altered as the authors became more familiar with the problems facing them, a reasonable development in all serious speculation. At first they had proposed that influenza viruses are themselves generated in an appropriate milieu in comets, and they explained the anomalous distribution of influenza in South

Wales schools by viral invasions from space. Microbiologists and others were not slow to object that influenza viruses are sophisticated parasites, closely tailored to the immune situations in their human and other hosts, and that almost every season they undergo the physicochemical changes that we have already described. Hoyle and colleagues accordingly modified their hypothesis and the 1986 version is both more complex and more plausible.

They now propose that the living particles that periodically arrive from space are not influenza viruses but that they are relatively undifferentiated precursors that they have named "viroids," a name already in scientific use for another life-form. These viroids are then tailored by the hosts' immune systems to become the influenza viruses that are familiar to virologists. The space-infected humans themselves suffer influenza and transmit the matured virus to their nonimmune companions by direct spread. The proportion of directly space infected to secondarily infected individuals is said to differ in different diseases. In influenza by far the greater proportion of cases is said to be attributable to space infection by viroids.

Microbiologists and epidemiologists with experience of the influenzal parasites in field and laboratory have found difficulty in accepting any necessity for assuming invasions from space, and it is a claim of the new concept proposed in this book that it can explain the behavior of influenza and its viruses without recourse to any such mechanism.

The speculations of Hoyle and his colleagues have, however, received much publicity in the press and on radio and television, and their semipopular books are widely read. The publicity has rendered a service by directing attention to the inadequacies of the current concept of influenzal epidemiology. Chapter 4 of their small book *Viruses from Space*[1] contains stimulating speculations about the nature of "viroids" and genes, owing much to the logic of computer science. They view the "viroid" as an address or a system of addresses permitting the host cell to search the large areas of its genome that are "hidden" and normally untranscribable. Such unexpressed DNA may be considered as sections of the genome to which the cell has lost or never possessed appropriate addresses. It is a hidden data bank that the cell cannot address and cannot normally use, and it contains "a cosmically determined store of evolutionary potential accumulated by cells through geological time." It is not made clear why this hidden store of data should have been cosmically determined.

The "viroid," it is suggested, provides the missing address(es) that allows the cell to operate this normally inaccessible part of its own genome, and they quote as experimental evidence the finding of Keilin and Wang that diseased leguminous plants sometimes produce hemoglobin. Apparently the DNA of legumes contains a normally unexpressed gene sequence for hemoglobin production and the disease provides the address for the plant to operate it. It is difficult to understand how the finding of Keilin and Wang supports the Hoyle hypothesis. It does not explain why "viroids" generated in cometary tails should contain addresses to hidden areas of the genome of human or other cells.

Their second hypothesis is that the addresses carried by an invading "viroid" initiate a "program modification" in human and other animal host cells that causes the genome to produce influenza viruses. They see the process as analogous to the production of hemoglobin by the diseased leguminous plant. But the odd result of leguminous plant disease is not analogous to their influenza hypothesis unless the vegetable disease has also been caused by "viroids" that were generated in comets. They claim experimental support for their second hypothesis because the influenza virus requires an activated cell nucleus before synthesis of viral RNA and protein can occur, and because inhibitors of cellular DNA also inhibit viral replication. But here again there seems to be no support for their central tenets. They are saying no more than that influenza viruses are obligate parasites of mammals and birds using the genetic machinery and indeed other parts of the host cell for replication and assembly. This is common knowledge and it is not clear how they adduce experimental support for their second hypothesis from it.

Their discussion about the mechanisms causing antigenic variation of influenza virus is interesting and suggestive, but it too links the changes with invasions from space of "viroids" carrying the addresses to hidden sequences of DNA. "Viroids" could only obtain the appropriate addresses by evolving in a situation exposed to the now hidden sequences of DNA. The authors do not suggest the existence of such a situation somewhere in space.

Their final hypothesis is that when the virus has been transcribed, it is capable of replication "in the usual way," spreads from cell to cell, and eventually causes an attack of influenza. They agree that it would "in principle" also spread directly to other persons, but this method is less common because free influenza virions are so vulnerable. Influenza cases caused by "viroid" infection from space are thus supposed to preponderate over the number of cases caused by person-to-person spread. They consider the opposite to be true in, for example, smallpox in which more cases result from direct spread than from primary space infections.

For several reasons we have given considerable attention to the space invader hypothesis. It deserves attention as standing almost alone in attempting to provide an overall picture of influenzal epidemiology and the authors have tackled large problems left unexplained by the current concept. Their ideas have commanded wide interest in the nonmedical public and in medical circles that do not specialize in influenza. If their hypotheses are incorrect, they deserve careful rebuttal. Hoyle's theory that life here did not originate on Earth and that primitive life forms are still reaching Earth from space is a fascinating speculation with important evolutionary consequences on which other scientists with the requisite expertise will doubtless comment. There are at least two reasons for rejecting the assumptions concerning influenza. First, the space "viroid" hypotheses do not evade the necessity for explaining the evolutionary history of influenza viruses in their numerous host species on Earth. The "viroids" must themselves have undergone an appropriate evolutionary history to be able to address the hidden sequences of the genomes in a variety of hosts, and the authors have made no attempt to describe the physicochemical nature of the putative address system. Second, a simple explanation is preferable to one that is more complex, and the new concept proposed in this book offers a simpler alternative. The only extraterrestrial influence postulated by the new concept is the seasonally mediated stimulus that recalls the virus to infectiousness from persistent or latent infection, and we can witness the operation of such seasonally mediated stimuli all around us in the seasonal behavior of countless other animals and plants.

A mathematical rebuttal of the hypothesis of Hoyle and Wickramasinghe is given by Henderson *et al.*[2]

A NEW METEOROLOGICAL HYPOTHESIS

An article that only came to notice in mid-September 1989 will be considered here because the explanation of influenzal epidemic behavior that it proposes differs from any of those examined in Chapter 16 and in other chapters. The hypothesis had been first proposed tentatively by Raddatz in September 1987 at a seminar in the University of Manitoba, Canada. It was published in detail by Hammond, Raddatz, and Gelskey[3] in the May–June 1989 issue of the *Reviews of Infectious Diseases.* They are suggesting that:

> . . . long-range atmospheric transport of aerosolized influenza virus may contribute to the spread and persistence of influenza virus and that seasonal changes in atmospheric circulation patterns and dispersive characteristics may lead to the regular annual cycles of influenza activity.

The authors possess both microbiological and meteorological expertise. They describe how aerosols are dispersed into the atmosphere mainly in two ways. First, they travel through the planetary boundary layer between 100 m and 1500 m above the Earth's surface and are dispersed horizontally by the wind for distances varying from ten kilometers to hundreds of kilometers. Second, they are at times hoisted to the upper atmosphere where they are conveyed much greater distances before returning to the Earth's surface, some of them having formed the nuclei of raindrops. The smaller particles may remain suspended for days or weeks before being returned by a downdraft. Dust from China travels more than 10,000 km across the North Pacific, smoke from forest fires in Canada sometimes reaches Europe, and the fallout of radioactive material from the nuclear explosion at Chernobyl in the Ukraine was detected worldwide.

Global dispersal of aerosols in the atmosphere is not random. For example, an atmospheric pathway links Asia with North America:

> In winter, the southeast coast of Asia is an active frontal zone with frequent cyclogenesis, i.e., development of low pressure centers. Surface aerosols, and potentially aerosolized influenza virus, may be conveyed to the upper levels of the atmosphere by these systems. There they usually encounter a fast westerly flow that transports them towards North America.[3]

The authors suggest that in winter the Far East may be a source whence the North American continent receives influenzal viral aerosols a week or two after their emission. In summer this atmospheric pathway weakens and alters direction, a change that they suggest may be responsible for the virtual absence of influenza from North America during the summer months.

The final dispersal of the aerosols after their transoceanic journey would be confined to the lower planetary boundary layer. Thus seasonal atmospheric conditions would contribute to the winter season prevalence of influenza.

The atmospheric pathway that they have described is not unique:

> . . . pollution studies have identified other source-to-sink pathways within the global circulation of the atmosphere. Ultimately all these pathways are interconnected. Thus the entire population of the world may become exposed to airborne influenza virus. This may help to explain the ubiquity and persistence of the disease.[3]

Aerosols containing infectious influenza virus particles are said to be ejected by the coughing and sneezing of patients suffering from influenza, and the plausibility of the hypothesis depends in part on the longevity of the infectiousness of the aerosolized particles. If infectiousness is maintained, airborne virus has been shown to be effective at a far smaller dose than the same sort of influenza virions administered intranasally, and in some experiments it has seemed possible that a dose approximating a single virus particle is able to initiate an attack of influenza in man. The conditions of temperature and humidity affecting aerosols within the atmosphere at all levels are considered to be compatible with persistence of influenza viral infectiousness even for the duration of long flights especially at high altitude where temperature is low.

The lethal effect of ultraviolet radiation on influenza virions, intense at high altitudes, receives little consideration.

Despite the modest claim of the authors that the hypothesis of long-range atmospheric transmission of influenza virus may in part explain some features of the epidemic behavior of influenza, they have made a bold attempt to elucidate the mechanisms operating the regular annual cycles of influenza activity in North America and, by implication, in all other parts of the world.

> It is postulated in this paper that atmospheric dispersion and intercontinental scale transport of aerosolized influenza virus may contribute to the spread, persistence and ubiquity of the disease, the explosiveness of epidemics, and the prompt region-wide occurrence of outbreaks and that seasonal changes in circulation patterns and the dispersive character of the atmosphere may help to explain the regular annual cycle of influenza activity.[3]

The claim is not unreasonable from their evidence of atmospheric carriage of pollutants and, if correct, would be a major contribution to understanding how the present concept of direct spread from the sick might be producing the behavior of epidemic influenza as it is experienced globally in the world population. There are, however, difficulties in accepting it. Naturally emitted aerosolized droplets from persons or animals suffering from influenza do not appear able to infect susceptible subjects at a distance of a few meters. It seems inconceivable that they should do so at a distance of many thousands of kilometers after a flight above the clouds, subjected for many hours to the lethal rays of the sun unfiltered by much of Earth's atmosphere.

Reference to the problem list in Chapter 18 (see section entitled "The Problem List as a Totality") shows that whereas the meteorological concept might be helpful in explaining problems 1, 2, 7, 8, and possibly 6 and 20, it makes no contribution toward the solution of the other 15 problems that are inexplicable by the current concept.

The meteorological hypothesis is not readily compatible with the new hypothesis advanced in this book unless symptomless carriers of reactivating influenza virus are emitting aerosolized virus during their infectious period.

INFORMATION FROM EVOLUTIONARY DENDROGRAMS

Influenza viruses base their replication on an RNA genome the instability of which we recognized as a major contributor to the continuing evolution of the virus (Chapter 5). It has recently become possible for molecular virologists to examine in detail the genetic changes that have taken place during antigenic drift and shift. The information that is accruing from such analyses is highly relevant to concepts of the epidemiology of influenza and the transmission of the viruses.

Fitch[4] and Fitch and Harris[5] have described methods for producing dendrograms, minimal-length evolutionary trees, by the analysis of differences in the number of changes in the nucleotide sequence of related strains isolated at different dates. Two examples will illustrate the use of these methods

Gillian Air[6] determined the nucleotide sequences for around 20% of the hemagglutinin gene of 32 strains of influenza A virus representing all the then known 12 hemagglutinin subtypes. She also predicted their amino acid sequences by using the genetic code. She comments:

> When the sequences of . . . three subtypes of HA (H2, H3 and H7) are compared, at either the nucleotide or amino acid level, they are remarkably dissimilar. The complete lack of cross-reactivity between them is well characterized but, because substitution of a single amino acid can completely destroy interaction between a particular antigenic determinant and monoclonal antibody and because there is a limit to the number of independent determinants on the hemagglutinin, the degree of divergence in sequence is remarkable. . . .

> Although the number of nucleotide changes within a subtype increase[s] with time, the vast majority of those are "silent" and do not alter the protein sequence except in the apparently less constrained signal peptides.

The author studied the matrix and nonstructural protein genes of human influenza A viruses of 1933–77 and she found a gradual accumulation of nucleotide change and some changes of amino acids. The rate of change of the H1 genes from 1934 to 1957 was much the same as that of the H2 gene from 1957 to 1968 and of the H3 gene from 1968 to 1977. The antigenic drift of H2 was continued in avian hosts after 1968 and was still doing so in 1977.

Air drew three conclusions from her findings:

1. Antigenic drift within the hemagglutinin subtypes of influenza A virus has progressed continuously at a rate of about 5% of nucleotide changes per 20 years.
2. Within each subtype of influenza A virus, the rate of sequence change in the hemagglutinin-coding gene is not significantly greater than that in the genes that are not subjected to antigenic selective pressure.
3. Antigenic drift in any HA subtype sequence has never been found to tend toward any other subtype.

She notes that if drift were to progress indefinitely, the sequence variation would in time outstrip the differences that exist between subtypes, but drift seems not to be the mechanism that has brought about these larger changes. Whereas the range of amino acid changes within subtypes varied only between 0 and 9%, that

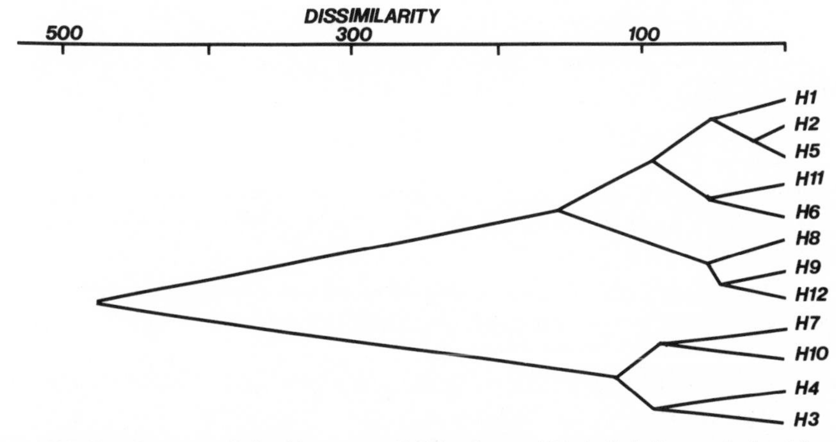

FIGURE 12.1. Sequence relationships among 12 HA subtypes. "Dissimilarity" was calculated from the amino acid identities, then the dendrogram was calculated from their relationships (from Air,[6] Fig. 2; reproduced with permission from *Proceedings of the National Academy of Sciences, USA*).

between the subtypes varied from 20 to 74%. Moreover, no intermediate strains have been found that link one subtype with another.

The dendogram, Figure 12.1, is the evolutionary tree based, by the author, on the progressive dissimilarity of the strains examined. She concludes her fine paper:

> The data presented here give no hint of the mechanisms underlying two extremes of virus evolution—i.e., how a 1950 H1N1 influenza virus reappeared almost unchanged in 1977 or how 12 subtypes that have little recognizable sequence homology have evolved.

Our second example was published nearly a decade later and ventures to put a date on the common ancestor of all the human hemagglutinin subtypes of influenza A virus (Gammelin *et al.,* 1990).[7] In order to discover the common ancestor of those viruses that have been evolving separately in humans and waterbirds, and also to discover the relationship between the nucleoproteins of influenza viruses belonging to types A, B, and C, the authors have undertaken a thorough comparison of sequences of the nucleoproteins of many strains.

They state that the nucleoprotein is a main determinant of the species specificity of influenza A viruses, and that the nucleoprotein (NP) genes seem to be solely responsible for preventing the mixture of the two large reservoirs of these viruses, one in humans and the other in waterbirds. Gammelin *et al.* describe their work and speculation as follows:

> Using 25 NP sequences we have constructed evolutionary trees by the strict-parsimony procedure of Fitch (1971). In contrast to the evolutionary-gene tree, the tree based on amino acid sequences unravels remarkable differences between avian and human NPs, differences which are best explained by a strong differential selection pressure on the human NPs. It is speculated that this selection pressure is caused by a change of the host and the (T-cell) immune response. A cautious extrapolation of the tree suggests that the human influenza A virus NPs evolved ~150 years ago from an avian ancestor.

Their study is also considered to illuminate the origin of the other types of influenza virus:

> . . . influenza B and C viruses . . . have a common root with influenza A viruses. Influenza B and—apart from a rare isolation in pigs—influenza C viruses were found only in humans. These viruses might have emerged also from an avian influenza A ancestor a correspondingly long time ago and, by the selection pressure specific for humans, might have developed finally their own type.

They surmise that NP genes may have been exchanging slowly between human and avian strains "for some time," and they refer to the special character of strains that belong to the human H1N1 subtype of influenza A virus. These mostly possess the ability to rescue avian genes (fowl plague virus–NP temperature-sensitive mutants) in chicken embryo cells, though they do so much less efficiently than the avian strains of A(H1N1) virus. The human A(H1N1) strains possess this property until the USSR/77 strain, whereas all later human A(H1N1) strains are unable to rescue avian genes. It will be remembered that the 1977 USSR strain was virtually identical with the 1950 Scandinavian A(H1N1) strain. Gammelin *et al.* comment:

> This would imply that, from time to time, whole branches of human influenza A virus NPs reconstitute their own type. The other viral genes . . . have to evolve correspondingly. A new NP gene then becomes reimplanted again from an animal reservoir, giving rise to a strain highly attenuated for humans with a lag phase in virulence reminiscent of the antigenic shift and slowly adapting to the new host.

The mechanism they are suggesting clearly does not operate for influenza B viruses that possess no known animal reservoir apart from mankind.

The most striking way in which the waterbird reservoir of influenza A virus differs from the human reservoir is in the location of the virus in the alimentary canal of the bird where it stimulates little or no immune reaction in its host. Readers are encouraged to study this most interesting paper. Figures 12.2 and 12.3 show the evolutionary tress that the authors have deduced from the nucleotide and amino acid changes.

One hundred and fifty years may seem a short space of time for the influenza A virus to have achieved diversification into three dissimilar subtypes in the human host species, but, when one considers the replicative potential of the virus, the instability of the RNA genome and the prodigious number of replications that occur in even a single influenza epidemic, such a result seems not unreasonable. Nevertheless, there are difficulties about accepting their explanation of the reappearance of A(H1N1) strains 20 years after their disappearance in 1957. Air could find no hint of the operative mechanism, but the available evidence suggests a mechanism that differs from that offered by Gammelin *et al.* as quoted above. Serological evidence indicates that the major serotype previously called A prime had had an era of prevalence not only from 1977 and 1946, but also from about 1908. The evidence also indicates that both the H2 and the H3 subtypes had had previous eras of prevalence during the last 100 years.

Gammelin *et al.* are correct in saying that their findings also "imply that, from time to time, whole branches of human influenza A virus NPs reconstitute their own type," but their subsequent hypothesis that "A new NP gene then becomes reimplanted again from an animal reservoir . . . " involves operations of formidable complexity unless the animal reservoir is human. The use of the word "reservoir" is convenient and evocative but carries a danger of oversimplfying the situation. The reappearances of each of the three human influenza A subtypes occurred worldwide within a single season, an achievement that seems impossible if the virus had to be reimplanted each time from a nonhuman animal reservoir.

The interspecies transmission of an influenza A virus resulting in its pandemicity must be a protracted affair probably requiring many false beginnings. We

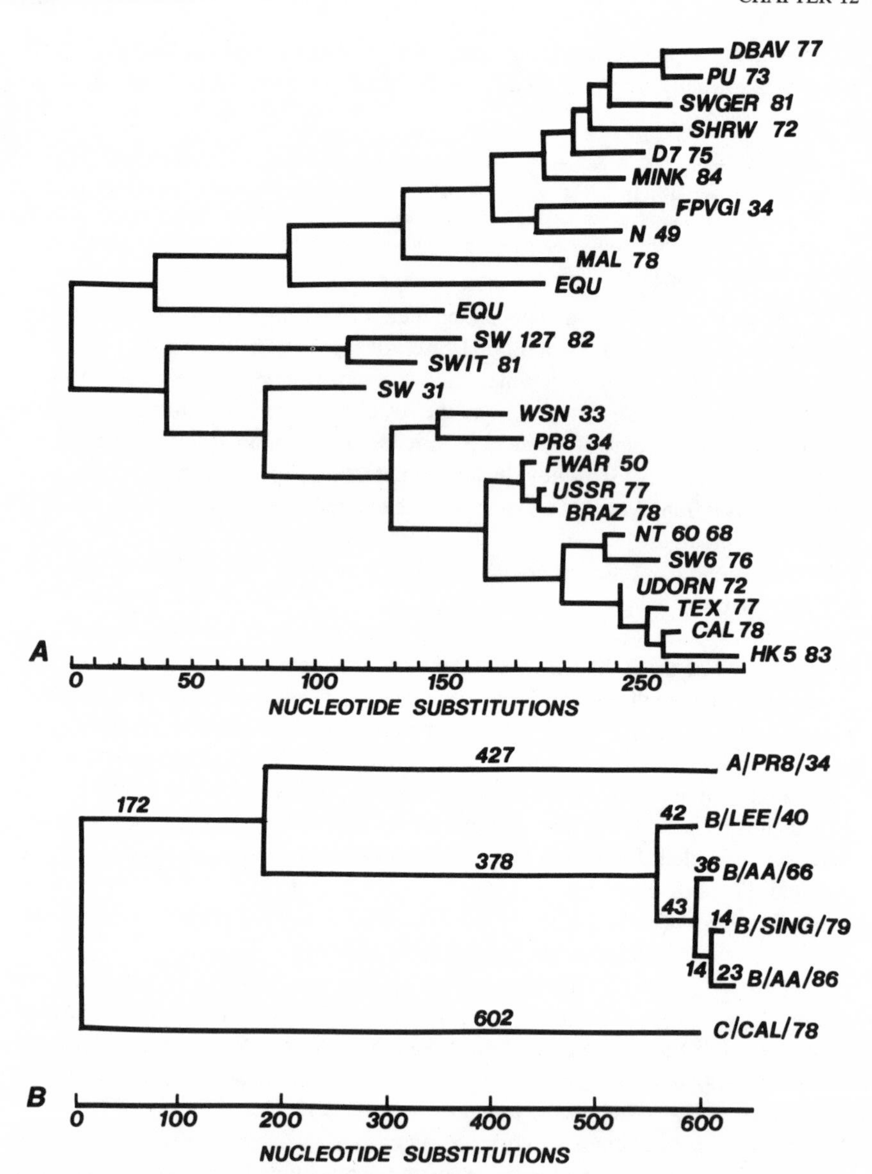

FIGURE 12.2. Phylogenetic trees of NP genes of influenza viruses. (A) Tree of 25 influenza A viruses derived from the coding region of the nucleotide sequences. (B) Tree of influenza A, B, and C viruses. Influenza A virus mutates at 3.3 nucleotide substitutions per year, and influenza B virus at 1.15 per year (from Dr. Christopher Scholtissek; reproduced with permission from *Molecular Biology and Evolution*).

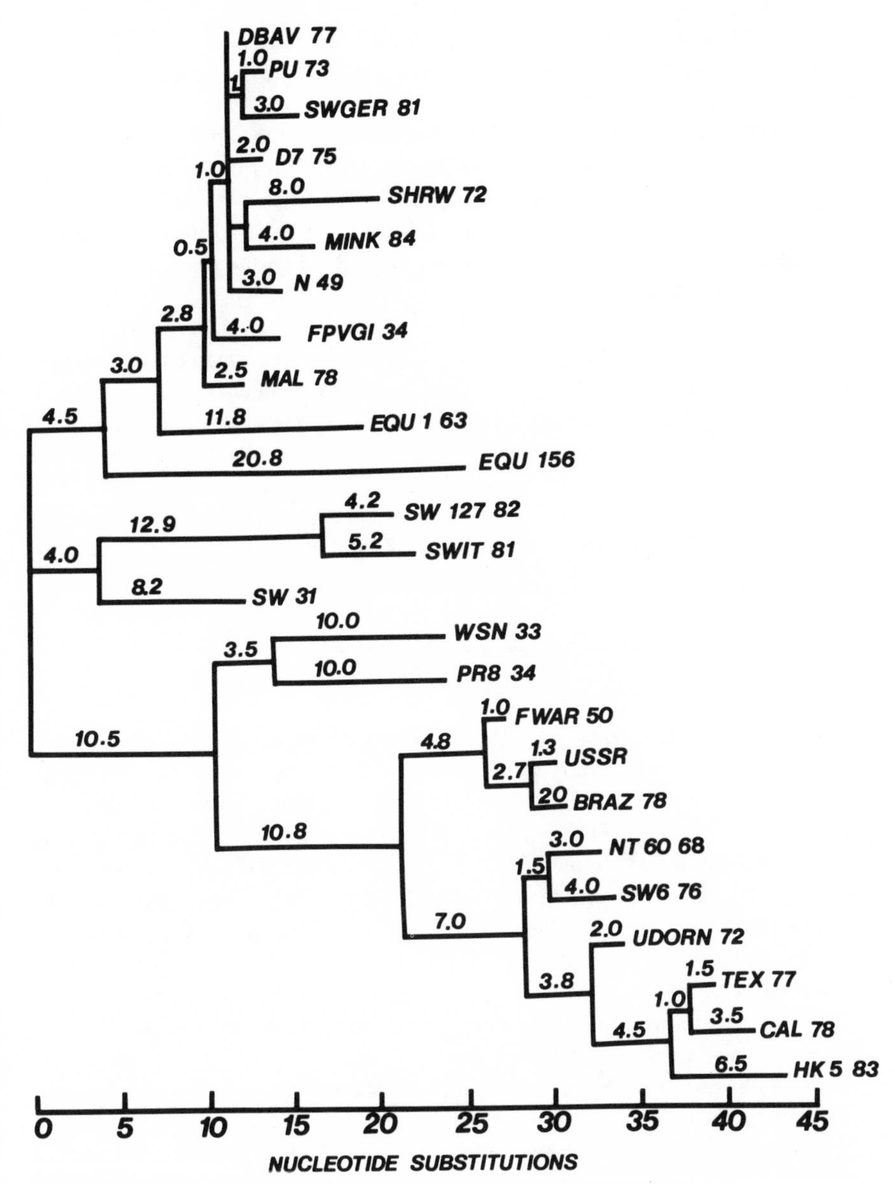

FIGURE 12.3. Most parsimonious tree for the amino acid sequence of influenza virus NPs from 25 type A strains, analyzed in terms of nucleotide substitutions by back-translating the changed amino acid into ambiguous codons (from Gammelin *et al.*,[7] Fig. 2; reproduced with permission from *Molecular Biology and Evolution*).

shall be discussing the problem in the following chapters. Here it is sufficient to say that the available evidence seems to suggest that transference of each of the major serotypes of influenza A virus, H1, H2, and H3, from waterbirds to mankind must have occurred long ago, perhaps centuries before the origin suggested by these authors. The most probable explanation of the worldwide reappearances seems to be that the genome had been stored in the tissues of human carriers who had suffered the infection. This hypothesis also possesses difficulties that are discussed in later chapters, but the worldwide availability of a reservoir of human carriers is an attractive speculation.

Although originally humans may have obtained all three subtypes from avian hosts, another possibility ought to be remembered. As Air pointed out, given time, antigenic drift might well have diversified the human influenza A virus antigenically until the strains differed sufficiently for genetic reassortment to produce the subtypes we now know. The H1N1 influenza A subtype may have been illustrating the process during the present century. Figure 10.1 showed how successive eras of prevalence of major H1 serotypes have succeeded one another since about 1908, one major mutant (now known as H1N1 old style) having had three eras.

POSITIVE DARWINIAN EVOLUTION IN HUMAN INFLUENZA A VIRUSES

Air's observation, that "the rate of sequence change in the hemagglutinin-coding gene is not significantly greater than that in the genes that are not subjected to antigenic selective pressure," has recently been contradicted by the findings of Fitch and colleagues (1991).[8] They had studied the aminoacid replacements at an antigenic site on the HA molecule of 20 strains of human H3 influenza A virus, each isolated in a different season from 1968 to 1987. They found that the evolutionary tree of the H-coding gene was unusual in having a long trunk with short lateral branches "like a cactus." The hemagglutinin of the viruses in the trunk (survivors) were fixing proportionately more aminoacid replacements than those in the branches (nonsurvivors), indicating "that type A human influenza virus is undergoing positive Darwinian evolution."

This conclusion is reinforced by their other findings, namely that evolution was three times as rapid on these H-coding genes as on nonstructural genes, and also that nonsurvivor strains lasted on average for only 1.6 years. Their extinction was both common and rapid.

The authors regard this behavior as the primary strategy whereby human influenza A virus, unlike most other viruses, attempts "to outrun its pursuers . . . a treadmill running to escape immune surveillance." The behavior is found in both human and equine influenza A infections but not in those of birds.

The findings have an important bearing on the subject of this book, the transmission of epidemic influenza. The evolution that is changing the nature of influenza A virus is clearly being driven by the pressure of the immune surveillance by its human host. The question therefore arises: at what point in the association between the virus and its host is the immune pressure being applied?

In section 6 of Chapter 9 reason was given for dismissing the possibility that immune selective pressure is being exerted by the host in the first days of his infection, the days when the virus is commonly supposed to be naturally transmitted. The recipient too, if he is to develop an attack of influenza, has no immunity that will exert selective pressure at the time of receiving the infecting dose. The evidence of positive Darwinian evolution of influenza A virus in human infections thus adds to the probability that the virus is being transmitted at a time when the human donor host has developed specific immunity from the attack of influenza so that the virus must then be acquiring the adaptations to evade it.

SOME ARBOVIRUSES THAT MAY BE RELATED TO INFLUENZA VIRUS

There are two groups of tick-borne viruses, Dhori and Thogoto viruses, that may be of interest in relation to the evolution of influenza virus.[9] The ticks are parasitic on certain herbivorous mammals such as deer, goats, sheep, cattle, and camels. The viruses cause severe febrile illnesses in the mammal but do not incommode the ticks and are passed vertically to their offspring. They resemble influenza virus in being enveloped virions containing a single-stranded RNA genome of negative sense. Like influenza C virus, the nucleoprotein core consists of only seven segments. Molecular virology should now be able to determine by their nucleotide sequences whether they are related to influenza virus, and, if so, whereabouts they should be included in the orthomyxovirus evolutionary tree.

REFERENCES

1. Hoyle F, Wickramasinghe C, Watkins J: *Viruses from Space and Related Matters.* Cardiff, University College of Cardiff Press, 1986.
2. Henderson IM, Hendy MD, Penny D: Influenza viruses, comets and the science of evolutionary trees. *J Theor Biol* 140:290–303, 1989.
3. Hammond GW, Raddatz RL, Gelskey DE: Impact of atmospheric dispersion and transport of viral aerosols on the epidemiology of influenza. *Rev Infect Dis* 11:494–497, 1989.
4. Fitch WM: Toward defining the course of evolution: Minimum change for a specific tree topology. *Syst Zool* 20:406–416, 1971.
5. Fitch WM, Harris JS: Evolutionary trees with minimum nucleotide replacements from amino acid sequences. *J Mol Evol* 3:263–278, 1974.

6. Air GM: Sequence relationships among the hemagglutinin genes of 12 subtypes of influenza A virus. *Proc Nat Acad Sci USA* 78:7639–7643, 1981.
7. Gammelin M, Altmüller A, Reinhardt U *et al:* Phylogenetic analysis of nucleoprotein suggests that human influenza A viruses emerged from a 19-century avian ancestor. *Mol Biol Evol* 7:194–200, 1990.
8. Fitch WM, Leiter ME, Li X *et al:* Positive Darwinian evolution in human influenza A viruses. *Proc Nat Acad Sci USA* 88:4270–4274, 1991.

13

Influenza in Nonhuman Hosts

THE IMPORTANCE OF INFORMATION CONCERNING NONHUMAN INFLUENZAL INFECTION

Most of us consider influenza as a human malady, an acute respiratory illness lasting for a few miserable days or at most a week or two before the invading virus is eliminated by the immune defenses of the body. We may be aware that some animals too are subject to influenza, horses for example, and we have heard of outbreaks of cat 'flu and dog 'flu.

Such a picture seriously underrates the global dominion of influenza. Influenza A viruses are widespread parasites of numerous species of avian and mammal host. There is a danger in transferring knowledge about the behavior of the virus in one species to explain how it behaves in a different sort of host because virus and host make specific adaptations to one another. Nevertheless, faced as we are with so many complexities of its relationship with mankind, it is valuable to know how it has managed to adapt itself to survive as a successful parasite in other species of host.

Only a few of the many species of mammals and birds now known to be subject to natural influenza A virus infection can be discussed, but those chosen exhibit a range of modes of parasitism that illustrates the versatility of the virus.

Influenza B virus is not in the same category as type A virus. In the many influenza B virus infections reported as occurring in birds and mammals, the virus has always been derived from mankind. It has not yet been found as a natural parasite of any nonhuman host species, nor to cause epizootics, and it is therefore considered to be a natural human parasite, entirely dependent on mankind for its survival.

THE ZOONOTIC HYPOTHESIS

The hypothesis that hosts other than man are somehow involved in the antigenic shift of human influenza A virus subtypes has been called zoonotic in contradistinction to that which proposes that the affair has been contained within the human host species for at least the last 100 years, the so-called *anthroponotic* hypothesis. The latter name has been frowned on by classical scholars, but it has crept into influenza literature, fills a need, and nicely balances the name of the opposing hypothesis, so both names will be used here.

The zoonotic hypothesis takes more than one form. Some proponents suggest that novel subtype strains preexist as complete viruses in an alternative host species, others suggest that only some genes (always including the gene coding for the hemagglutinin) are obtained by genetic reassortment of the human virus with a virus from a nonhuman host. The possibility that both processes may occur from time to time is also entertained.

The requirements are as follows: the appropriate viruses must exist in nonhuman hosts, must either have the opportunity of infecting a human host or alternatively of encountering and genetically recombining with human influenza A virus, and the recombinant must subsequently have opportunity to invade a human being.

That is not all. Evidence is needed that such interspecies transmission to man is actually occurring, and that it can be succeeded by immediate adaptation of the novel strain to the human host, and further that the new strain can infect mankind worldwide within a few months and continue an era of world prevalence for a decade or more, often having very rapidly displaced the predecessor subtype from the world.

Some of these requirements are satisfied. Nonhuman hosts can harbor human influenza A viruses. A(H1N1) and A(H3N2) human subtypes have passed naturally from man to swine and have become adapted to their new host species.[1] Back-transfer from swine to man occurs much less commonly but has been reported for strains of both subtypes, and we shall be describing later in this chapter a considerable human outbreak caused by a swine influenza virus. We have already mentioned the opportunity for genetic reassortment presented by the close association of domestic ducks, swine, and humans in some parts of the world (see the section "Hypothesis of an Influenza Epicenter in China" in Chapter 11). A number of observers including Masurel and his colleagues,[1] and Kilbourne[2] and Andrewes[3] have pointed the finger at domestic swine as the prime suspect for providing an accessible reservoir of old and newer strains of influenza A virus from which humans can receive strains when suitable opportunity occurs.

Before discussing the arguments on both sides of the debate between proponents of the zoonotic and anthroponotic hypotheses, it will be well to take a brief look at the wider parasitism by influenza A virus.

AVIAN INFLUENZAL INFECTIONS

For many years fowl plague has been recognized as a highly lethal pestilence in flocks of chickens and turkeys. As long ago as 1900, the causal agent was found to be a filter-passing organism, but its influenzal nature was not established until 1955. Since then many species of bird have been found to be subject to infection by numerous influenza A virus subtypes. Wild waterfowl are host to so many varieties that they have been proposed as the hosts of the primal influenza A virus from which all other host species including mankind have derived their influenzal parasites.[4]

The huge avian reservoir of influenza A virus remained undiscovered for a quarter of a century after Shope had discovered the swine influenza virus because microbiologists were hunting for it in the wrong site. Influenced by their experience in man and pig, they were seeking to isolate the virus from the respiratory tract, but in wild and domestic waterfowl it lodges in the gut, and, moreover, infected ducks are seldom ill.

The influenzal parasitism of waterfowl therefore differs greatly from that of man, mammals, and many other sorts of bird. In discussing the hypothesis of an influenzal epicenter in China (Chapter 11), we saw how the virus, multiplying harmlessly in the intestine of domestic and wild ducks, is excreted in their droppings and infects lakeside mud and pond water where the birds congregate. Ducklings that have been infected by contaminated water may or may not develop a transient rise of temperature and a humoral antibody response that wanes as the bird matures. The virus, however, remains for a long time harmlessly replicating in the gut, and the bird tolerates the simultaneous presence of several subtypes. Since all 30 known combinations of H- and N-coding genes have been found in ducks, their intestine provides a situation ideal for producing reassortant viruses. It is not surprising that waterfowl are strongly suspected of being somehow implicated in the mechanism of antigenic shift of human influenza A viruses.

Avian influenza is not always so innocent, especially when the virus infects birds other than waterfowl. In some outbreaks of fowl plague the virus, A(H7N7), kills nearly all the chickens or turkeys in the affected flocks, and it is found to be widely disseminated in all organs of the dead birds.

Other influenza A viruses may also cause avian epizootics. In April 1983, chicken flocks in Pennsylvania began to be plagued by a widespread outbreak of influenza A caused by A(H5N2) strains. For several months the disease was mild, but suddenly in October the virus became more pathogenic, killing a high proportion of the infected birds. When virulent isolates were compared with their less virulent predecessors, they were found to have been derived by a minimal adaptation. There had been a change in one amino acid near the stalk of the hemagglutinin spikes and in another near one of the antigenic sites on the spikes. These small changes had been associated with the disappearance of the defective interfering particles that had apparently been moderating the pathogenic capability of the virus.[5]

We shall be discussing later a possible role of defective interfering particles in the epidemiology of human influenza. Here we must note that avian influenzal infections provide us with evidence of several modes of host–parasite relationship that can be evolved by the influenza virus:

1. Acute severe infections with blood-borne distribution of the virions throughout the avian body and an ability to cause 100% mortality.
2. The possibility to switch from low to high pathogenicity by a minor mutation on the gene coding for the hemagglutinin.
3. The capability to adapt to prolonged innocuous parasitism in the alimentary canal.
4. Symptomless temporary infection.

Severe pathogenicity in avian influenza is commonly a property of viruses that contain either H5 or H7 hemagglutinin antigen. The mild H5N2 virus in Pennsylvanian chickens between April and October was probably the abnormal strain.

SWINE INFLUENZA AND OTHER INFLUENZAL INFECTIONS IN PIGS

The porcine epizootics known as swine influenza or hog 'flu are reputed to have first appeared in pig farms in Iowa in October 1918 when the great autumn wave of the human influenza pandemic was rising to its maximum. The pig farmers and the veterinarians claimed that it was a new disease among pigs. In 1935, Patrick Laidlaw[6] at Hampstead speculated that the human influenza virus had in 1918 infected the swine in Iowa, and that it had become adapted to the new host species and established itself as an enzootic parasite of pigs, which thereafter became subject to the epizootics that are called swine influenza. He pointed out that the accuracy of his speculation could be tested by antibody studies, and he was proved to have been correct when it was found that persons who had suffered type A influenza between 1918 and 1929 often possessed humoral antibody to Shope's swine influenza virus, later designated A(Hswine1N1). The virus presumed to have caused the human influenza A from 1918 to 1929 therefore came to be known as the A(Hswine1N1-like) strain, and the period of its human prevalence "the swine era." Both those human and swine strains are now included in the A(H1N1) subtype along with the human A(H0N1) and A(H1N1 old style) strains.

Swine influenza resembles the human disease in that the virus cannot commonly be isolated from individual pigs between the seasonal epizootics in the herds. Shope[7] considered that the virus was surviving by a complex parasitic cycle involving both the pig lungworm and the earthworm as intermediate hosts. The lungworm, a common parasite of pigs, has a strange life cycle. The mature worms pass from the lung up the bronchi and windpipe into the throat and are swallowed. They lay their eggs in great numbers in the gut of the pig, so that the eggs are dropped with the pig's feces. When eggs are swallowed by earthworms they hatch into larval lungworms. Infested earthworms are devoured by pigs rootling for food, and the lungworm larvae pass through the intestinal wall of the pig, enter its bloodstream, and so reach the lung where they settle, become mature, and repeat the cycle.

Shope considered that he had evidence that swine influenza virus had hitched onto the pig–lungworm–earthworm cycle, becoming latent in mature lungworms, passing vertically into their eggs, and so into the earthworm and thence again into the pig's lung with the lungworm larvae. An additional seasonally mediated stimulus was invoked to activate the virus from latency and precipitate the porcine influenzal attack. He also incriminated a close relative of Pfeiffer's influenza bacillus called *Haemophilus influenzae suis* as an additional precipitant.

In 1960, Sen, Kelley, Underdahl, and Young[8] confirmed Shope's hypothesis, using pathogen-free colostrum-deprived pigs obtained by hysterotomy and raised in individual isolation units, and fed a sterilized diet. They claimed to have reactivated the latent virus by multiple injections of extract of another worm parasitic in pigs, the nematode *Ascaris* whose larvae migrate through the lungs. The latent virus was also found to be reactivable throughout the year whereas Shope could only obtain reactivation during autumn and winter, the seasons during which swine influenza occurs naturally.

Such experiments showed that swine influenza virus could be transmitted by way of the pig–lungworm cycle, but it was later found that epizootics of the disease occur as commonly in herds that are free of lungworms as in lungworm infested herds. G.D. Wallace[9] considers that the pigs themselves are harboring influenza virus in some mode of latency from one epizootic to the next, probably in the respiratory tract. He, like Shope, suggests that human epidemic influenza may be operating by a similar mechanism. Wallace concludes that, though Shope's pig–lungworm cycle may be able to operate in swine, it is not a necessity for the epizootiology of swine influenza in pig farms.

Swine influenza caused by A(Hswine1N1) strains has spread gradually since 1918 to herds of pigs in most parts of the world. In 1949, Young and Underdahl[10] suggested that the virus might be causing abortions and stillbirths in pregnant sows. In 1962, Menšik[11] in Hungary produced evidence that the virus was being carried asymptomatically in breeder sows and was being transmitted transplacentally to their piglets, findings that were confirmed ten years later by Nakamura *et al.*[12] The suppositions of Young and Underdahl were verified in 1982 when Gourreau and colleagues[13] isolated A(Hswine1N1) strains from spontaneous abortions and stillborn piglets of dams that had suffered natural swine influenza during seasonal epizootics of the disease.

Gourreau *et al.* later found that transplacental transmission of influenza virus by swine is not confined to A(Hswine1N1) strains of A(H1N1) subtype, but also that human A(H3N2) strains spreading naturally in herds of pigs could be isolated from abortions and stillbirths of infected sows.

The porcine fetus, however, is not necessarily killed by the transplacental influenza virus but may survive to be born with a symptomless and apparently harmless influenza virus parasitism, which endures for much of the early life of the pig.

The similarities between influenzal parasitism in the two host species, man and pig, deserve consideration because the seasonal outbreaks in both species may have a similar mechanism. Easterday[14] agrees with Wallace[9] that an unidentified stimulus dependent on season must be precipitating epizootics of influenza in herds of swine by reactivating influenza virus from latency in some carrier pigs. The hypothesis is similar to that concerning human epidemics proposed by the new concept, although it is not clear whether such porcine carriers fall ill when transmitting reactivated virus, or whether they usually remain symptomless as proposed for human carriers by the new concept.

The acute influenzal illness in pigs and the epizootic in herds of swine resemble the human illness and the epidemic in mankind. On the other hand, vertical transplacental transmission causing death of the conceptus or symptomless infection of the infant seems not to occur in human influenza, or if it does so the phenomenon is uncommon or unrecognized. Human abortions and stillbirths are seldom attributed to influenza.

We have already seen how readily domestic pigs receive human strains both of A(H1N1 old style) and of A(H3N2) influenza virus, and how rapidly the human strains become enzootic and epizootic in swine.[15] Pigs can also receive influenzal infection from such other hosts as domestic ducks so that within the pig the influenza A viruses are able to exchange genetic information by reassortment and recombination of their genes. The situation is a basis for speculation about a zoonotic mechanism by which mankind can acquire or has acquired major antigenic varieties of influenza A virus. Porcine influenzal transmissions to man do occur but they do so much less readily than transmissions in the opposite direction. Only once has such a transfer from swine to man been reported as causing a considerable human outbreak, namely at Fort Dix in 1976, and the porcine virus has not yet been known to establish itself as endemic in mankind. The Fort Dix outbreak is of such interest and importance that it merits a section of this chapter.

Study of influenza in the pig allows us to add to the list of modes of influenzal infection:

1. Transplacental transmission causing death of conceptus
2. Transplacental transmission causing inapparent infection
3. Latency of the virus in the pregnant sow and in the piglet

THE FORT DIX INFLUENZA EPIDEMIC

Early in January 1976, as the soldiers stationed at Fort Dix, New Jersey returned from their Christmas holiday, an intake of recruits from many parts of North America arrived with them. Almost immediately an epidemic of influenza broke out which was at first thought to be part of the general outbreak of A(H3N2) influenza currently attacking the United States.[16] Throat washings collected on 29 January from influenza patients admitted into the Fort Dix hospital did in fact yield seven isolates of A(H3N2) strains, but four of the specimens, one of which had been obtained from a 19-year-old recruit who died from his illness, yielded isolates of a different influenza virus. This was subsequently identified as a strain of A(H1N1) subtype similar to the A(Hswine1N1) strains that had been causing influenza in swine since 1918.

U.S. epidemiologists were faced with a serious decision because the swine influenza virus is deemed to be the homologue of the virus that caused the awesome human influenza in 1918. Was the Fort Dix outbreak, which had already attacked more than 100 military personnel and killed one healthy youth, the forerunner of the return of the 1918 pandemic 'flu? They advised the president to order a program of general vaccination with the virus derived from humans who had been infected by the swine influenza virus.

It was a difficult assignment. Despite strenuous efforts only five strains of A(Hswine1N1) virus were isolated although there was serological evidence that at least 230 recruits had been infected by it. The swine-like strain was far more difficult to isolate than the contemporary A(H3N2) strains. A prodigious effort by the pharmaceutical companies provided sufficient vaccine. Microbiologists in the United Kingdom advised against comprehensive vaccination.[17] The Fort Dix outbreak did not spread and it remains the only large outbreak of human influenza caused by virus transmitted from a nonhuman host to have been recorded.

This zoonotic human outbreak demonstrated that it is possible for an animal influenza virus to become adapted to parasitism in mankind, and the interest that it generated brought to light a number of incidents in which persons associated with pigs developed influenza caused by the swine virus. The transfer appears to be not uncommon as an isolated event, but adaptation to man to cause outbreaks spreading within the human community is rare. The swine virus was probably originally a human parasite, so it is surprising that renewed adaptation to cause human epidemic influenza occurs so infrequently, the disease now being so common in herds of swine in most parts of the world.

EQUINE INFLUENZA

In past centuries the contemporary accounts of human influenza epidemics have frequently mentioned outbreaks of a similar disease that were occurring around the same time in horses. Here is an example in a description of the influenza in Edinburgh during the autumn of 1732:

> We believe it will not be improper here to mention, the horses in and about this place being universally attacked with a running of the nose and coughs towards the end of October and the beginning of November, before the appearance of this fever of cold among men.[18]

Dr. Robert Whytt of Edinburgh mentioned another equine epizootic of influenza in Scotland in a letter to Sir John Pringle. An influenza epidemic had begun in some schoolchildren around 20 September 1758. He continues:

> A gentleman told me, that in the Carse of Gowrie [which is a valley north of the river Tay in Perthshire, Scotland] in the month of September before this disease was perceived the horses were observed to be more than usually affected with a cold and a cough.[19]

In answer to the questions about animal influenza in Dr. John Fothergill's questionnaire about the 1775 epidemic, Dr. William Cuming of Dorchester replied that he had heard from a reliable witness that a disorder had prevailed very generally among the horses in Yorkshire in August, a month or two before the arrival of human influenza.[20]

Dr. Haygarth[21] of Chester supported Dr. Cuming's account of an equine epizootic around that time. His medical correspondent in North Wales had written that all the horses thereabouts had been seized with coughs about August and September.

Dr. Thomas Glass[22] wrote that around Exeter also both horses and dogs were severely affected in September with colds and coughs. Fothergill himself had noted that in London:

> During this time, the horses and dogs were much affected; those especially that were well-kept. The horses had severe coughs, were hot, forbore eating, and were long in recovering. Not many of them died that I heard of; but several dogs.[23]

A general impression was current that the animal disease possessed some sort of continuity with the human epidemics. The error was pardonable. Equine influenza resembles the human disease in its manifestations and behavior. It arrives dramatically causing simultaneous illness in many horses, donkeys, and mules in a locality. Before the invention of the internal combustion engine when these animals provided the main transport and means of human communication, equine influenza epizootics could cause devastating dislocation to a community. A local newspaper gives a vivid account of such an emergency when an epizootic among the horses in 1872 in Louisville, Kentucky brought the city to a standstill:

> The disease involved was equine influenza that swept down from Canada in November, virtually immobilizing horse-drawn Louisville. . . . Horses were in great distress with mucus from the nostrils, coughing and running at eyes. . . . Over 1000 sick animals [horses and mules] withdrawn from service. Stoppage of passenger vehicles of every kind. Patent leather and hobnails splash mud together. Citizens called on to aid the fire department. . . . The city's physicians had to walk to visit their human patients and undertakers couldn't bury anybody because there were no horses do draw the hearse. . . . At least dangers of reckless driving in the streets have ceased.[24]

Although few horses died, the community became so imbued with talk about the disease that "epizootic" became a local household word to describe family ailments.[24]

The equine disease resembles human influenza but, if correctly reported, it differs in that the epizootics are not seasonal. Outbreaks tend to occur when many horses from widely disparate sources congregate as at race meetings, a situation favoring either direct spread or spread by healthy carriers. Caution is needed in accepting the significance of the nonseasonal nature of equine epizootics. Much of the evidence comes from experience with racehorses, and it may be biased both by the newsworthiness of epizootics at race meetings and by the fact that horses may be assembled from great distances. According to the new concept, normal seasonal patterns may be breached by transporting reactivating carriers across zones of latitude. Out-of-season epidemics may be explained by such mobility in man, and the erratic timing of the 1918 pandemic in the southern hemisphere has been attributed to the postwar movement of immense numbers of men and women. Gerber has said that latent equine influenza virus may explain the apparent absence of the virus between epizootics in horses.

Two influenza A subtypes, H7N7 and H3N8, cause equine epizootics. There is little cross protection between them, so when both sorts of the virus are causing contemporaneous epizootics, some individual horses, mules, or donkeys may suffer attacks by both subtypes within a short time.

The early observers were presumably mistaken in believing that continuity existed between equine influenza epizootics and human influenza epidemics. Despite the age-old intimate relationship, no such sharing of influenzal parasites occurs as that which characterizes the influenzal infections of man and pig. Nevertheless, at some period there may have been a direct or indirect exchange of genetic material because the H3 hemagglutinin of the human A(H3N2) strains that appeared in 1968 in the first outbreak of "Hong Kong influenza" was found to be similar to that of an equine strain that had been isolated in 1961.

INTERSPECIES TRANSMISSIONS OF INFLUENZA VIRUS

The early writers were also mistaken about the continuity of human influenza with canine illness. Both human A and B strains are occasionally transmitted to dogs from human influenzal patients but they do not cause illness in the dog and have never been reported as initiating a canine epizootic. So what was the disease in dogs that looked like canine influenza? The most likely cause of confusion was probably canine distemper, which causes an influenza-like illness. We have already seen how the canine distemper virus, because it accidentally infected laboratory ferrets, delayed identification of the first isolation of human influenza virus in 1933.

"Cat 'flu" too is not caused by an influenza virus, but by other agents. As with dogs, cats are sometimes harmlessly infected by a human strain of influenza virus.

When interspecies transmission of influenza virus takes place, the parasite often behaves in a different manner in the new host. If it is correct that the virus causing the 1918 pandemic was the progenitor of swine influenza virus, it caused a milder disease in the pig than in man and it was more antigenically stable in the new host. The human A(H3N2) strains too cause milder illness in swine and undergo less antigenic variation during enzootic parasitism, and they are transplacentally transmitted in breeding sows to cause abortion, stillbirth, or latently infected piglets.[13]

Birds are the best interspecies transmitters of influenza A viruses. The transmissions occur more readily between different species of bird than from birds to mammals. Water birds most readily transmit their influenza virus between species such as ducks, gulls, terns, and puffins, and they suffer less severely from influenzal illness than such land birds as chickens, turkeys, parrots, and budgerigars. The colonial habit of many water birds combines with their defecation into the ambient water to ensure that they have been sharing their influenza viruses for so long that hosts and parasites have evolved a friendly *modus vivendi* not yet achieved by land birds.

Difference in body temperature is suspected as being an important barrier to interspecies transmissions and may partially protect mammals from avian influenza viruses habituated to a higher body temperature. The protection is far from complete as was dramatically illustrated recently. In 1979, the harbor seals *Phoca vitulina* on the northeast coast of the United States began dying of pneumonia and it was estimated that 20% of the harbor seal population was killed by the epizootic. The cause was found to have been an avian influenza virus related to fowl plague virus, A(H7N7), previously never isolated except from birds.[25]

A few years later, in the 1982–83 season, a different avian influenza A virus caused an epizootic that killed a large number of New England harbor seals. Before these experiences no influenza virus isolated from mammals had been able to replicate in the gut of chickens, but both these seal viruses did so, thus confirming that two avian influenza viruses had caused natural epizootics in seals. The seal mortality far exceeded the worst recorded human influenzal mortality.[26]

Dr. A.S. Beare, who reported on strains sent to England from the Fort Dix epidemic, has long been interested in interspecies transmission of influenza A viruses. He makes the following comments in a letter dated 27 August 1987:

> In regard to your observations on the relationship between human and animal type A viruses, I thought you might like a résumé of my own experience (and that of Rudi Kasel) on human infectivity/virulence of various wild-type influenza viruses
>
> *Equine influenza viruses.* (1) A/equine/1/56 (H7N7) is totally noninfectious for man. Reassortants of this virus containing the H7 HA and NA + other assorted genes of virulent human strains are also noninfectious for man. (Beare, unpublished material; Beare, *Prog. Med. Virol.* 1975, 20, 49–83). (2) A/equine/Miami/1/63 (H3N8). Produces modified influenza in man (Kasel *et al., Nature,* 1965, 206, 41–43).
>
> *Swine influenza viruses.* A/swine/Taiwan/7310/70 (H3N2) readily infects man but is not as virulent as prototype A/Hong Kong/1/68 (H3N2). Earlier viruses with the conventional antigens H1N1 (swine) seem to have low (but definite) human infectivity (Beare *et al., Lancet,* 1971, 1, 305–308).
>
> *Human viruses* (various serotypes) vary widely in their virulence for volunteers but are always liable to induce clinical influenza (Beare and Craig, *Lancet,* 1976, 2, 4–5).
>
> *Avian viruses.* Serotypes tested were H1N1 (swine), H3N8, H3N2, H6N2, H6N1, H9N2, H4N8, and H10N7. They all have very low infectivity, there was no virus-shedding but a few anti-H3 rises. A/duck/Ukraine/1/63 (H3N8), said to be the precursor of A/Hong Kong/68 (and of A/equine/63) was no more infectious than other serotypes but it was not easy to find seronegatives. However, it was not the same as equi/63. Dare I say that avian influenza is irrelevant to the human disease?

When considering the possible involvement of nonhuman hosts in the epidemiology of human influenza, three different results of interspecies transmission need to be distinguished. The first is the transfer of influenza virus from an individual of one species to an *individual* of another species, a not uncommon phenomenon sometimes resulting in illness of the new host. The second is transfer that causes an *outbreak* by spread of the virus between individuals of the new host species. Examples of this much rarer result are the limited outbreak of swine influenza in recruits at Fort Dix in 1976 and the severe outbreaks of avian influenza in harbor seals in 1979 and 1982. Third, transfer may result in *permanent adaptation* of the influenza parasite to the new host species as in 1918 when a human influenza A virus appears to have established itself successfully as an enzootic parasite of the pig, the new host species being thereafter subject to seasonal epizootic influenza. A similar result seems to have followed transmission of A(H3N2) strains from humans to swine because, in addition to widespread enzootic parasitism in herds of pigs, A(H3N2) epizootics have already been reported. The relationship between the influenza viruses in these two hosts may be peculiar because, although the enzootic virus in the pig remains antigenically rather stable, the human minor variant A(H3N2) strains are acquired by pigs almost as soon as antigenic drift in man occurs. The same relationship is also occurring with the human A(H1N1 old style) strains that reappeared for an era of human prevalence in 1977. These have also been acquired by swine.

In order to evaluate the attempts to explain antigenic shift in human influenza A viruses by genetic reassortment with those of nonhuman hosts, we need to know not only how readily the human virus can be transmitted to alternative hosts and vice versa, but also how readily humans can be infected by viruses of human origin that have undergone genetic substitution when encountering nonhuman strains in other species of host. It is not enough to know their potentiality for establishing themselves throughout the world population, and, if they can do so, the mechanisms whereby they achieve their new parasitism.

Human influenza viruses commonly pass to animal hosts, and hemagglutinins analogous to the H1, H2 and H3 of human Type A strains have been found in the influenza viruses of waterbirds, chickens, bats, whales, squirrels, deer and most domestic pets. Pig handlers are occasionally infected from pigs, but other reports of this opposite trend are scarce.

Kawaoka, Krauss, and Webster[27] have explored the evolutionary pathways of a polymerase gene PB1 of influenza A virus and discuss the light that this sheds on the interspecies transmissions of the virus and its individual genes. The evolutionary tree that they have constructed from the nucleotide sequences suggested that the PB1 gene of A/Singapore/1/57 (H2N2), the virus that caused the Asian pandemic of 1957, probably came from an avian influenza virus and was maintained in humans until 1968. However, the PB1 gene of A/NZ/60/68 (H3N2), the virus that replaced it worldwide in the human pandemic of 1968, had been derived from a different avian influenza A virus and it was still present in human influenza A virus strains in 1988. The avian PB1 gene had also been introduced into domestic swine.

Alignment of the deduced amino acid sequences of the PB1 genes of influenza A viruses isolated from different host species showed that all were able to encode a polypeptide of 757 amino acids. The amino acid sequences possessed a high degree of homology among all the influenza A PB1 genes. Some of the conserved regions of the PB1 gene of influenza A virus were found to be conserved in the corresponding regions of influenza B virus.

The findings of this study also support the hypothesis that pigs received their swine influenza virus from humans during the 1918–19 influenza pandemic.

When considering the implications of their findings the authors state:

> The avian PB1 genes and genes encoding surface glycoproteins [H- and N-] appear to have been introduced into humans in both the 1957 and 1968 pandemics. An association between avian-to-human transmission of surface glycoprotein genes is understandable because such genes confer selective advantages to influenza virions under pressure from immune systems.[27]

They consider that had the interspecies transmission occurred only once it might have been a random association of PB1 and HA-encoding genes, but that repeated interspecies gene transmission may indicate a preferential association between these genes. When the authors write of explaining how the PB1 genes were transferred from avian species to humans in both 1957 and 1968, it is not clear if they consider that the actual genetic reassortments occurred in those years. The new concept accepts their explanation of how the reassortments were produced, but maintains that the process could not have taken place in those years. The epidemiological evidence seems overwhelming that the reassortments must have happened many years or centuries earlier, and the inactivated and stored reassortments must have been reactivated about 1957 and 1968.

Discussions on herd immunity in Chapter 9 and elsewhere are relevant to their remarks on the influence of immune pressure on antigenic variation in human influenza A virus.

A recent investigation by Mandler *et al.*[28] found that marine mammals (whales and seals) from both Pacific and Atlantic oceans must have derived their influenza viruses from seagulls and mallards.

Beare and Webster[29] have recently found that 11 of 40 volunteers inoculated with avian subtypes of influenza A (namely subtypes H4N8, H6N1 and H10N7) shed virus and had mild symptoms but produced no detectable antibody response. The authors consider that virus multiplication was insufficient to stimulate a detectable immune response. Nevertheless, their results demonstrate that novel HA genes may be able to invade the reservoir of human influenza A genes by reassortment between avian and human viruses.

REFERENCES

1. Masurel N, de Boer GF, Anke WJJ *et al:* Prevalence of influenza viruses A-H1N1 and A-H3N2 in swine in the Netherlands. *Comp Immunol Microbiol Infect Dis* 6:141–149, 1983.
2. Kilbourne ED: Epidemiology of influenza, in Kilbourne ED (ed): *The Influenza Viruses and Influenza.* New York, Academic Press, 1975, pp 483–538.
3. Andrewes CH: Influenza A in ferrets, mice and pigs, in Stuart-Harris CH, Potter CW (eds): *The Molecular Virology and Epidemiology of Influenza.* London, Academic Press, 1984, pp 1–3.
4. Alexander DJ: Avian influenza viruses—recent developments. *Vet Bull* 52:341–359, 1982.
5. Bean WJ, Kawaoka Y, Wood JM *et al:* Characterization of virulent and avirulent A/chicken/Pennsylvania/83 viruses: Potential role of defective interfering RNAs in nature. *J Virol* 54:151–160, 1985.
6. Laidlaw PP: Epidemic influenza: A virus disease. *Lancet* 1:1118–1124, 1935.
7. Shope RE: Influenza, history, epidemiology and speculation. *Publ Health Rep* 76:165–178, 1958.
8. Sen HG, Kelley GW, Underdahl NR *et al:* Transmission of swine influenza virus by lungworm migration. *J Exp Med* 112:517–520, 1961.
9. Wallace GD: Swine influenza and lungworms. *J Infect Dis* 135:490–492, 1977.
10. Young GA, Underdahl NR: Swine influenza as a possible factor in suckling pig mortalities. I. Seasonal occurrence in adult swine as indicated by hemagglutinin inhibitors in serum. *Cornell Vet* 39:105–119. 1949.
11. Menšik J: Experimental infection of pregnant sows with *Influenza suis* virus. I. Proof of virus in placental tissue and in organs of newborn piglets. *Vedecke Prache-Vyzkumngho Ustavu Veterinarniho Lekanstrvi v Urne* 2:31–47, 1962.

12. Nakamura RM, Easterday BC, Pawlisch R *et al:* Swine influenza; epizoötiological and serological studies. *Bull WHO* 47:481–487, 1972.
13. Gourreau JM, Kaiser C, Madec F *et al:* Passage du virus grippale par la voie transplacentaire chez le porc dans les conditions naturelles. *Ann Inst Pasteur* (Virol) 136E:55–63, 1985.
14. Easterday BC: Animal influenza, in Kilbourne E D (ed): *The Influenza Viruses and Influenza.* New York, Academic Press, 1975, pp 464–488.
15. Wallace GD: Natural history of influenza in swine in Hawaii: Prevalence of infection with A/HK/68 (H3N2) subtype virus and its variants 1974–1977. *Am J Vet Res* 40:1165–1168, 1979.
16. Top FH, Russell PK: Swine influenza A at Fort Dix, New Jersey (Jan–Feb 1976). *J Infect Dis* 136:376–380, 1979.
17. Beare AS, Craig JW: Virulence for man of a human influenza A virus antigenically similar to "classical swine viruses." *Lancet* 2:4–5, 1976.
18. Medical Essays and Observations, published by a Society in Edinburgh, in Thompson T (ed): *Annals of Influenza in Great Britain from 1510–1837.* London, The Sydenham Society, 1852, p 42.
19. Whytt Robert: in Thompson T (ed): *Annals of Influenza in Great Britain from 1510–1837.* London, The Sydenham Society, 1852, p 63.
20. Cuming, William: in Thompson T (ed): *Annals of Influenza in Great Britain from 1510–1837.* London, The Sydenham Society, 1852, p 94.
21. Haygarth John: in Thompson T (ed): *Annals of Influenza in Great Britain from 1510–1837.* London, The Sydenham Society, 1852, p 111.
22. Glass Thomas: in Thompson T (ed): *Annals of Influenza in Great Britain from 1510–1837.* London, The Sydenham Society, 1852, p 102.
23. Fothergill J: in Thompson T (ed): *Annals of Influenza in Great Britain from 1510–1837.* London, The Sydenham Society, 1852, p 89.
24. Coady JH: The 1872 epizoötic. *J Am Vet Med Assoc* 170:668, 1977.
25. Webster RG, Hinshaw VS, Naeve CW *et al:* Pandemics and animal influenza, in Stuart-Harris CH, Potter CW (eds): *The Molecular Virology and Epidemiology of Influenza.* London, Academic Press, 1984, pp 40–41.
26. Kilbourne ED: *Influenza.* New York, Plenum, 1987, p 244.
27. Kawaoka Y, Krauss S, Webster RG: Avian-to-human transmission of the PB1 gene of influenza A viruses in the 1957 and 1968 pandemics. *J Virol* 63:4603–4608, 1989.
28. Mandler J, Gorman OT, Ludwig S *et al:* Derivation of the nucleoproteins (NP) of influenza A viruses isolated from marine mammals. *Virology* 176:255–261, 1990.
29. Beare AS, Webster RG: Replication of avian influenza viruses in humans. *Arch Virol* 119:37–42, 1991.

14

Experimental Studies

LABORATORY ANIMAL STUDIES REVEAL VARIED MODES OF SURVIVAL OF INFLUENZA A VIRUS

We have hitherto considered the varied modes of parasitism and host–parasite relationship that have evolved naturally between influenza virus and a number of different species of host. Laboratory studies are contributing much to our understanding of these processes, but only a brief account of a few of them can be given.

The following letter, dated 13 November 1987, was received from Sir Christopher Andrewes, then aged 91:

> Many years ago I planned to study the epidemiology of mouse 'flu. The mice I inoculated died of pneumonia, but their contacts all remained normal. An American worker had found that 'flu would go by contact in mice, so I repeated his work as exactly as I could, using the same strain of mice and virus; same result [i.e., virus did not spread]. I thought you'd be intrigued. I'm not sure if the virus goes in ferrets.

In 1983, Jakab, Astry and Warr in the United States[1] found that, in mice suffering from pneumonia caused by direct infection with human influenza virus, the infectious virus disappeared within nine days, but the viral antigen persisted in the alveolar cells of the lung in high concentration for more than a year, a long period in the life of a mouse.

In Moscow, in 1981, Frolov, Shcherbinskaya, and Gavrilov[2] had shown that a persistent infection by a human A(H3N2) strain could be produced in mice, and that infectious virus could be regularly isolated for 45 days and periodically thereafter for nine months. Changes in the antigenic profile of the virus were found to have occurred after five months, and pathogenicity diminished or was lost after seven months. After the first two months some of the isolates were already differing from the parent strain in their thermal stabilty.

The duration of infection has been found to vary with the type of challenge. In 1981, V.A. Zuev and colleagues[3] attempted to model latent influenza virus infection in mice in three different ways: (1) by causing influenzal illness, (2) infection with live virus vaccine strains, and (3) by vertical transmission to offspring of virus persisting in pregnant female mice. They found that, after recovery from influenzal illness, infection persisted for 112 days from the date of infection. After a single exposure to live vaccine the virus persisted for only 35 days. The situation was found to be very different in mice born from dams that were influenza virus carriers. The offspring were found to be carrying persistent influenza virus in high titer in their blood and viscera.

The same team followed the fate of such transplacentally infected mice. They had already noted that in directly infected mice persistent influenza virus could be detected only in lung tissue, whereas the vertically infected offspring of persistently infected dams suffered a general infection. This did not cause abortion or alter the duration of pregnancy, but after three weeks or more the transplacentally infected mice developed slowly progressive lesions in certain parts of the brain, the immune system, and the endocrine system, sometimes leading to death. They explain their findings as caused by a slow influenza virus infection, comparable with human congenital rubella or congenital lymphocytic choriomeningitis in mice, the immature status of the immune system of the fetus permitting transplacental transfer of virus leading to hypothalamic infection in the brain.

These workers also found that the baby mice were infected transplacentally even when their dams had been infected with influenza A virus before the babies had been conceived. Infectious virus could be isolated from blood, lungs, liver, kidneys, spleen, and brain of the offspring.[4]

Among many other experiments that demonstrate the versatility of the virus are the following:

Robinson, Easterday, and Tumova in 1979[5] attempted to reactivate influenza A virus that was latent in turkeys by stressing the birds in different ways. The virus could not be reactivated by the stress of heat or cold, nor by that of thirst or hunger, but the stress induced by crowded transport provoked a prolonged increase in the hemagglutinin-inhibiting antibody content of their blood, an indication that latent virus had been reactivated.

In 1978, Smolensky and colleagues,[6] studying the problem of virus carriers in chicken influenza, found that whole virus persisted for only 30 days whereas the hemagglutinin and neuraminidase antigens were present for twice as long. Moreover, administration of hydrocortisone on day 50 to such chickens provoked the appearance of several strains unrelated to the original virus. This last finding is reminiscent of that of Frolov and his colleagues[7] who reported that mouse spleens yielded A(H0N1) strains after two months of being infected with A(H3N2) influenza virus.

Such observations have often been dismissed as possibly caused by laboratory contamination, and they should therefore be repeated. We shall see, however, that similar claims have been made about persistent infection of cell cultures by influenza A virus.

Even leaving aside such subtype leaps, these findings have indicated various modes of parasitism available to influenza viruses that should be remembered during attempts to explain their natural relationship with the human host species. The virus has been shown to be capable of protracted infection during which it undergoes antigenic and other changes of varying degree; transplacental infection occurs and can cause a "slow virus" type of illness; viral antigen may persist in the lung long after infectious virus has disappeared, but infectious virions may be recalled by hydrocortisone, showing that at least the viral genome had been retained in some mode of latency.

PROTRACTED INFECTION OF CELL CULTURES

Cell culture offers a simpler system than the intact animal in which to examine the relationship of the parasite to the host cell. The chicken embryo in the fertilized egg was one of the earliest systems used to isolate viruses, and it is convenient to include it in this section although it is technically an animal (avian) host.

Many years ago, Burnet pointed out that the species of host from which the cell culture is derived may exert a selective effect on the influenza virions replicating therein. If so, we are not isolating from cell cultures the strain that actually caused the human disease. Schild and Oxford[8] and Patterson[9] among others have recently shown how the host cell exerts an evolutionary pressure toward antigenic variation on the parasitic virus comparable to that exerted by antibody in the intact animal. They showed that a human strain grown in chicken embryo differs from the same strain grown in mammalian cells. Possibly neither is identical with the parent strain that caused the human influenza. It is a sobering thought that almost all the laboratory study of human influenza has hitherto been carried out on viruses that differed from those that caused the human illness. Those grown in mammalian cells are probably closer to the human strains than those grown in avian cell culture or chick embryo. It has recently been shown that influenza A virus harvested directly from human infections seems identical with the same virus cultured in nonhuman mammalian cells.[10]

Persistent infection of the host cell occurs when a balance is achieved between replication of the virus and the normal functions of the cell. It can be initiated in a number of ways. Not long after the human influenza viruses were discovered, von Magnus[11] found that chicken embryos, if heavily infected with influenza virus, produced both normal (standard) infectious virus and nonpathogenic virions known at first as von Magnus particles. The latter were found to be deficient in matrix protein and are now known as defective interfering particles because they interfere with the replication of standard virions. At first considered to be aberrant, they are now known to occur in infections caused by many other viruses, and they play an important role in epidemiology and epizootiology.[12]Appearing early in the course of influenza, they moderate the severity of the disease, especially when the inoculum has been heavy. They do not themselves cause influenza, and in the last chapter (Chapter 13) we saw how a small mutation that appears to have operated against their production transformed a mild chicken influenza into a severe epizootic that killed up to 80% of affected birds.

The moderating effect of defective interfering particles on the replication of standard virions can be decisive, one defective particle being sufficient to inhibit the reproduction of large numbers of standard virions. Thus the infected cell, if it contains these noncytopathic defective particles, is able to come to terms with its parasitic colony, and a persistent infection is initiated that may endure for a long time.

The situation within persistently infected cells is not, however, stable. The remaining standard virions may be driven to evolve mutants that evade the interference of the defective particles, and their escape puts an evolutionary drive on the defective particles to respond, so that within the infected cell the evolution of the influenzal RNA genome may be rapidly proceeding.[13] One of the propositions of the new concept is that defective interfering particles produced early in human influenza induce persistent infection and interfere with direct transmission of the virus, and that the persistent infection drives the evolution of the virus and so is a part of the mechanism of antigenic drift of human influenza virus. Defective interfering particles are often used in the laboratory to initiate persistent influenza virus infection of cell cultures.

Fazekas de St. Groth was an early student of the antigenic changes in the virus during prolonged cultivation, first in Burnet's laboratory at the Walter and Eliza Hall Institute in Melbourne, Australia, and later in Europe where he was joined by Claude Hannoun.[14] The virus underwent antigenic drift in cell culture, and in this artificial situation they were able to isolate previously prevalent strains that had disappeared in Nature, and also new strains one of which subsequently appeared in the natural course of human epidemic influenza. They hoped to be able to produce such prophetic strains for inclusion in vaccines but antigenic drift proved to be too unpredictable (see also Chapter 9: Laboratory Production of Antigenic Drift of Influenza Virus).

In 1975, Daniel B. Golubev reported similar remarkable results obtained at the All-Union Influenza Institute in Leningrad.[15] A strain of A/Hong Kong/1/68 (H3N2) influenza virus in persistent culture drifted forward to strains identical with A/Victoria/3/72, which had appeared four years later, and shifted backward to strains resembling A/Singapore/1/57 (H2N2), belonging to a subtype that had vanished before 1968. St. Groth had made a similar claim for a subtype leap in persistent culture but was unable to substantiate it.

If such claims are confirmed, they must be showing that the genetic information needed for constructing strains of earlier subtypes is somehow conserved in virions belonging to later subtypes, and can be recovered during protracted infection of the cell. They are in keeping with the early serological findings of Henle and Lief described in "The Doctrine of Original Antigenic Sin . . . " in Chapter 6.

Golubev and Medvedeva[16] succeeded in keeping influenza A virus in persistent infection of cell culture of human embryo kidney or human embryo lung for periods varying from 40 to 289 days. They isolated 102 influenza virus strains of which they examined 44 in detail. The original antigenic profile of the hemagglutinin and the neuraminidase had been preserved in 31 of the 44, but was found to be markedly and permanently changed in the remaining 13.

In 1972, Gavrilov in Moscow[17] seems to have been the first to show that, in cell cultures maintained for up to 146 days, the influenza virus undergoes irregular cycles in which cytopathogenicity alternates with periods of quiescence and cell regeneration. The virus that he obtained from the culture fluid between days 27 and 105 replicated poorly and failed to agglutinate red blood cells. In one of his five cell cultures the hemagglutinin and cytopathogenicity returned spontaneously during the last days of the culture. In virus from two of the other cell lines these properties returned after serial passages in other systems.

Kantorovich-Prokudina and her colleagues[18] reported in 1980 that in prolonged culture, defective interfering particles alternated with standard virions for a period, but thereafter only the defective particles were produced.

In 1980, Barun K. De and Debi P. Nayak at the University of California–Los Angeles School of Medicine[12] showed that persistent infection of cell cultures could be initiated by co-infecting cells with both defective interfering particles and temperature-sensitive mutants (ts^-). Up to passage 7 such persistently infected cultures underwent cycles of cell lysis and virus production (crisis). After passage 20 they produced little or no virus, but they were resistant to reinfection by homologous virus although they could be infected by a heterologous influenza virus. Most persistently infected cells contained the complete viral genome, expressed viral antigens and produced infectious centers. Using cultures of avian, bovine, and human cells, the authors were able to show that the defective particles produced by any subtype would interfere with the replication of all influenza A subtypes. The production of defective interfering particles requires the presence of standard virions, and they found that this helper function operates whether or no the standard virion belongs to the same subtype. The maintenance of the cell cultures in a state of persistent infection depended on selection of a slow-growing temperature-sensitive variant rather than on the presence of a defective interfering virus or cellular production of interferon. Since most of the cells in a persistently infected culture contain and express viral antigens, the repeated crises in the earliest passages assist in the selection of a virus population with a behavior that permits continued cell viability. The defective particles seem to be particularly valuable in the early stages of establishing persistence by protecting the cells from the lethal potentiality of infection by standard virions and so allowing time for a less cytopathic variant to be selected. Once this has been accomplished, the defective interfering particles are no longer needed and may be eliminated.

The authors end their valuable paper thus:

> In nature as well as in cell culture there appear to be a continuous evolution of influenza viruses. DI viruses which appear commonly in influenza virus replication, may further aid in the selection of variants and thus help in the evolution of the virus and the creation of diversity among the virus population.[12]

Evidence is being found that viral genetic information is often stored for many years within the cells of the intact host. Physicians have long been aware that the paramyxovirus that causes measles may, very rarely, later cause subacute sclerosing panencephalitis, having been stored in the brain cells since the acute attack of measles. In 1984, Koch, Neubert, and Hofschneider[19] found that the RNA of the genes of a paramyxovirus remained lifelong in the brain of mice that had been infected. They detected its presence in brain tissue by means of a cloned genomic complementary DNA probe after all other evidence of the presence of the virus had disappeared. The latent viral RNA was not perceptible because it was expressing no proteins.

Other examples of genome storage of paramyxoviruses could be quoted, but they are not orthomyxoviruses, though related to them. In 1982, however, Stelmakh, Medvedeva, and Golubev[20] had shown that influenza A virus in protracted culture might behave similarly. When cytopathogenicity had ceased and virus could no longer be isolated, the viral RNA that had been synthesized during persistent infection continued to be produced. Complementary RNA and viral RNA were both being synthesized after the infectious virus could no longer be isolated.

These are among many observations that testify to the varied modes of host–parasite relationship available to influenza virus at the cellular level, and suggest that persistent infection and latency in more than one mode may be playing an important role in its natural behavior in mankind as in other host species. At present there is no microbiological evidence that influenza viruses undergo a latent stage of human parasitism, and the absence of such evidence challenges the authenticity of the new concept while in no way supporting the current belief in direct transmission.

Defective interfering particles are now known to be produced early during natural influenza in man so that the hypothesis of their interfering with direct transmission from the sick and initiating a period of persistent infection is not inherently improbable.

The awkward situation in which we find ourselves may be summarized as follows: The current concept of direct measles-like spread of human influenza cannot be correct because it leaves most of the epidemic behavior unexplained. The new concept explains most, possibly all, of the difficulties, but microbiological evidence has so far failed to establish the assumptions concerning latency and persistent infection in the human host. Should persistent infection be shown to occur naturally after human influenza the finding will add support to the new concept.

THEORETICAL APPLICATION OF EXPERIMENTAL FINDINGS

At least three modes of influenzal parasitism have been observed in laboratory investigations:

1. Acute cytopathic infection
2. Persistent infection induced by defective interfering particles and sustained by temperature-sensitive mutants
3. Latency of the viral genome

These three are together able to provide mechanisms that can explain the behavior of human influenza, during and between epidemics and eras of prevalence of major serotypes.

The acute infection is the mode that causes human illness by cytopathic damage as the virus spreads rapidly from cell to cell within the respiratory tract. The high multiplicity replication rapidly produces the defective particles that interfere with spread to other human hosts and initiates persistent infection lasting for months or a year or two as an innocuous symptomless carrier state in the recovered influenza patient. The new concept adds the proposal that a seasonally mediated reversion to infectiousness transforms the noninfectious carrier into an infectious focus around whom nonimmune companions are exposed to the opportunity to develop influenza. The virus causing their influenza is likely to differ from that which caused influenza in the carrier because, as explained in Chapter 9, the carrier will by then be specifically immune and often unable to transmit the original strain, and his companions choose the fittest of the mutant variants that he is shedding. Years later at the end of the era of prevalence of the major serotype, the persistent infection in all the carriers will have ceased, and the latent genome of the virus in them preserves the potential for renewing standard virions identical with those that had caused influenza in the carrier.

It seems likely that persistent infection usually persists only for a year or two, long enough to cause a subsequent wave of epidemic influenza in the world population. By contrast the latent genome may remain in the carrier for the rest of his life. With this view, carriers of the prevalent strain are widely seeded throughout the world population for the duration of the era of the major serotype, and latent genomic material is similarly widely seeded during the lifetimes of all those who suffered an attack of that sort of influenza. Many years after the end of the era of prevalence, when one or more generations of mankind have elapsed and the world population once again contains a high proportion of persons not immune to that major serotype, a reactivation of the genome causes the antigenic shift leading to a renewed era of world prevalence. The actual identity of the reactivated strain might depend on the year of the reactivation, most of those reactivated in any particular year being identical. Thus in 1977, latent genetic material seems to have been reactivated only in carriers who had suffered influenza A in the 1950–51 and 1953 epidemics, and specifically in those who had been infected by the Scandinavian strain.

Evidence for the correctness of this hypothesis is far from complete, but the behavior of epidemic influenza and its causal viruses and such experimental evidence as is available is thought to accord better with this than with other concepts hitherto proposed.

In 1979, J.J. Holland and his colleagues in California, working with vesicular stomatitis virus of cattle, found that antigenic changes were occurring more rapidly during persistent infection than during acute infection. Rapid and continuous evolution was taking place in the defective interfering particles in persistent infection whereas not much change was occurring in the standard virions during acute infections. Their conclusions are fascinating, reminiscent of Charles Darwin's observations on the evolution of the finches in the Galapagos Islands:

> The sequestered intracellular environment of persistently infected cells favors rapid and continuous evolution. . . . Persistently infected cells offer an ideal environment for virus genome evolution because persistently infected cells survive indefinitely and can allow a variety of virus mutants to arise and compete (and complement each other) for long periods without any need to mature and spread to other cells. Obviously, the less virulent mutants would be selected since more virulent virus will result in cell death and be eliminated from the surviving cell population, . . . these findings may have a significance for virus evolution in general. These data predict, for example, that a given strain of Type A influenza virus should tend to be genetically rather stable while spreading in acute infections during a pandemic. The extensive genome segment mutations which give rise to new variants (with or without reassortment) of influenza virus, however, may accumulate most efficiently in foci of persistent influenza within individual animals.[13]

They are beautifully describing how persistent infection can act as a factory of antigenic variation. Although they do not envisage the new concept, their description accords with it. The human carriers of persistent influenza virus would be exerting selective pressure at the cellular level on their persistent viruses during the long period of noninfectious carriage, even before reactivation exposed the various antigenic strains to the antibody in the host's circulation before the carrier shed influenza virus for transmission. Dr. John Skehel tells me that he cannot agree that persistently infected cells survive indefinitely.

ZOONOTIC *VERSUS* ANTHROPONOTIC CARRIAGE

The zoonotic hypothesis of antigenic shift proposes that human pandemic strains are being preserved between their eras of human prevalence in animal host(s) until recycled in mankind when a hiatus in the human immune situation provides a suitable opportunity for reinvasion of the human hosts.

In favor of the hypothesis is the evidence that the three human influenza A virus subtypes (as currently defined) possess different genes coding for hemagglutinin, and that in one of them the neuraminidase-coding gene also differs from that in the other two subtypes. Such genetic differences have probably come about by genetic reassortment, a process that readily occurs when strains of two or more different subtypes simultaneously co-infect the same host. As we have seen, waterfowl provide an ideal host, able to provide influenza A virus genes coding for the two surface proteins of all known subtypes and the pig provides a host in which avian and human strains are able to meet and exchange genes.

In 1968, when the Asian A(H2N2) subtype was replaced by the Hong Kong A(H3N2) subtype, it was found that the novel H3 hemagglutinin closely resembled that obtained in 1963 from both a duck in the Ukraine and a horse in Miami. The duck hemagglutinin differed in only 23 amino acid sequences from that in the human virus of five years later. The resemblance is close enough to bespeak a common ancestry for the hemagglutinin-coding gene in the three viruses.

Similarly, the N1 neuraminidase in the A(H0N1) and A(H1N1 old style) major variants was found to be similar to that in the swine influenza virus and also to that in an avian strain isolated from a duck in Germany in 1968.

A third similarity was found between the H2 hemagglutinin of the 1957 Asian strain in mankind and that of an avian strain isolated from a turkey in Massachusetts.

Such instances of relatedness must carry due weight when assessing the zoonotic hypothesis of the origin of human influenza A subtypes, and more instances are likely to be discovered. If the zoonotic hypothesis is correct, it is perhaps surprising that many more similarities have not yet come to light. They do, however, strongly support the view of Kilbourne (Chapter 11: Later Concepts of Kilbourne), Holland, and others on the potentiality of segments of influenza virus (and other) RNA to undergo evolutionary change independently of the virus package.

There can be no doubt that interspecies transmission of influenza A virus is occurring in Nature, and that the resulting association of dissimilar subtypes in a single host often permits an exchange of genes. If the genes exchanged include those coding for the surface protein hemagglutinin, the genetic reassortment results in the natural production of a strain differing in subtype from both parents. Such may well have been the origin of the three human subtypes, but the zoonotic process fails to explain their recurrent eras of human prevalence with long interpandemic absence.

ANTHROPONOTIC *VERSUS* ZOONOTIC CARRIAGE

If we accept the evidence provided by serological archaeology that influenza A virus subtypes are recycled in the human host species, then the first major problem to be addressed concerns the site and mode of storage of each subtype during the period of its absence between successive eras of prevalence. The evidence does not support the earlier concept that at each antigenic shift a strain totally novel to parasitism in mankind is being introduced.

Some indication would be valuable of the mode of parasitism adopted by the stored virus that we are seeking and of where it is most likely to be found. The absent strain is sometimes stored virtually unchanged for half a century, and neither the standard infectious virions nor their defective interfering particles meet our requirement because both are continually subject to antigenic changes caused by both mutation and genetic reassortment. The viral RNA, however, might be stored in an inactive condition not subject to mutation nor readily liable to genetic recombination. The viral genome would therefore be the most likely mode of retention of the potentiality to reproduce the parasite relatively unchanged after years of storage.

Where is the most likely place for it to be stored? In laboratory mammals infected with influenza virus, standard virions disappear from the cells of the respiratory tract almost as soon as the animal recovers from its illness, but viral RNA remains there for many months thereafter. The human respiratory tract would therefore seem to be the first place to seek for stored genomic material from which virus strains virtually identical with the strain that had caused the attack of influenza might have been destined to be reactivated many years later. If the genome is to be stored and reproduced, the most economical method would be to do so in the human host that suffered the illness.

We have been considering the problem in relation to storage and reproduction of subtypes, but we must not forget that the same phenomenon has characterized the major varieties that have arisen by mutation, not by genetic reassortment, namely A(Hswine1N1), A(H0N1), and A(H1N1 old style) strains.

The second problem confronting us is the means whereby at each antigenic shift or equivalent major antigenic variation the novel strain achieves world distribution within a single season. Here again clues are not altogether lacking. Antigenic drift caused by point mutations often behaves similarly. Vast populations occupying a large proportion of the Earth's surface suffered epidemic influenza caused by A/Port Chalmers/73 (H3N2) strains in the 1974–75 season. In 1975–76 season, the nonimmune persons in some communities suffered an influenza epidemic caused by the variant A/Victoria/3/75 (H3N2). These minor changes occurring in most influenza seasons make us familiar with the phenomenon of the rapid wide distribution of novel strains, and it is reasonable to speculate that a similar mechanism must be at work on a still larger scale at antigenic shift. Both shift and drift occur after an interepidemic period. In both, the predecessor strains usually disappear.

Communities isolated from the main streams of human communications would be expected at times to escape outbreaks of influenza suffered by the rest of the world population. They would fail to receive the immunity conferred by the epidemics and so would become immunologically out of step. This was exemplified when remote Amerindian communities in the Amazonian jungle suffered epidemic influenza caused by A(H2N2) strains that had ended their era of general world prevalence several years earlier (Chapter 15: Anachronistic Influenza Epidemics in Remote Communities). Such experiences show that vanished subtype strains are still available to the human community if an ecological niche is available. Monto and Maassab[21] found evidence of the continuing presence of past subtype strains in the human population.

The third problem facing the anthroponotic and zoonotic concepts is the vanishing trick: the disappearance of A(H1N1) strains in 1957 with the appearance of A(H2N2) strains in the Asian influenza pandemic, and the disappearance of Asian strains in 1968 at the appearance of A(H3N2) strains in the Hong Kong influenza pandemic. These were antigenic shifts, but we must not forget that the same vanishing trick characterized the major drifts that brought about the changes from A(Hswine1N1) to A(H0N1) in 1929, and then to A(H1N1 old style) in 1946. The vanishing trick has elicited wondering comment from many observers but no satisfactory explanation. Here again the analogy with what commonly occurs at the minor antigenic drifts provides a clue to a possible mechanism. An identical vanishing trick involving a smaller population characterizes most antigenic drifts. The new concept explains the vanishing of the predecessor at antigenic drift by a simple metamorphosis whereby the parent strain is arrested by the immunity of the donor who is a carrier infected in an earlier season. The recipients choose the fittest of the mutants shed by the carrier, so that the concept explains both the drift and the vanishing trick (see Chapter 9: Serious Difficulties in Explaining Antigenic Drift).

The vanishing of the predecessor strains at the major drifts that led to eras of world prevalence of A(H0N1) and A(H1N1 old style) must have been caused by the same mechanisms as minor antigenic drifts so that it seems likely that a similar mechanism is operating to cause the vanishing trick at antigenic shift. All these major antigenic changes probably came about originally many decades or centuries ago by a much slower zoonotic process. What we are seeing now is a recycling of those major variants; we are not really witnessing the antigenic changes.

At antigenic shift of subtype and at a major antigenic variation we are seeing the temporary disappearance of a family of strains that has effectively immunized almost all available persons in the world population who had been nonimmune to that subtype. That family of strains has filled its ecological niche and must await recruitment by births of a fresh population of susceptible human subjects.

Where should we be looking for the human carriers who are suggested as storing the genomes of vanished subtypes and of the mutant major variants, namely, the RNAs coding for the antigens of Hsw1, H0, H1 old style, H2, and H3? Two candidate groups suggest themselves:

1. Group 1: Those persons who suffered influenza during the era of prevalence of that subtype and remain immune to it. If the subtype has been long absent, many such persons will have died, others may have lost their latent virus, and the remainder will be elderly carriers. When the opportune moment arrives, their colonies may be reactivated to renewed infectiousness in seasonal sequence across the latitudes of the Earth by the seasonally mediated stimulus.
2. Group 2: An alternative suspect group must be the children born to women who suffered influenza caused by a strain of that subtype. On analogy with infected sows and mouse dams, the pregnant women may have been able to transmit the virus to their offspring even if their attack of influenza had preceded conception of the child. Table 14.1 shows the age range of the cohort of children conceived during each era of prevalence at the onset of subsequent prevalences of the same virus.

It may not be easy to determine whether the babies born, for example, during the first A(H2N2) era received the virus transplacentally and conserved it until the

TABLE 14.1. Age Group of Cohorts, Conceived during Eras of Prevalence of Major Serotypes, during Subsequent Prevalences of the Same Major Serotypes

Eras of prevalence	Serotype A(H2N2)	A(H3N2)	A(H1N1 O.S.)[a]
Era 1	1889–1900	1900–18	1908–18
Age group	0–11 yr	0–18 yr	0–10 yr
Era 2	1957–68	1968–88+	1946–57
Age group i	57–69 years	50–88+ yr	28–49 yr
Age group ii	0–11 yr	0–20+ yr	0–11 yr
Era 3			1977–88+
Age group i			59–80+ yr
Age group ii			20–42+ yr
Age group iii			0–11+ yr

[a]O.S. = old style.

next era, or whether they received that strain as their first influenzal infection after they had been born. An attractive suggestion is that only the first ever infection by influenza A virus of a person's life establishes a permanent intracellular genome of that strain. If correct, the hypothesis would unite both of the above suggestions. It is mentioned again in proposition 10 in Chapter 16.

There is no evidence as yet that either of the suspect groups contains carriers of influenza A virus RNA. Both groups are distributed worldwide and would be well-placed to distribute reactivated virus ubiquitously in a single season.

THE EVIDENCE FROM RETROSPECTIVE SEROLOGY

We have already described how sera collected from aged persons before the arrival of A(H2N2) strains in 1957 and A(H3N2) strains in 1968 possessed antibodies to these strains. The ages of such persons suggested that each had lived during a previous era of prevalence beginning about 1889 and 1900, respectively. In 1977, the recurrence of A(H1N1 old style) strains that had been absent for 20 years confirmed virologically the correctness of the concept that in mankind major serotypes of influenza A virus can be recycled with little antigenic alteration.

REFERENCES

1. Jakab GJ, Astry CL, Warr GA: Alveolitis induced by influenza virus. *Am Rev Respir Dis* 128:730–738, 1983.
2. Frolov AF, Shcherbinskaya AM, Gavrilov SV: Mechanism of persistent infection of influenza virus in the organism. *Abstract of the Vth International Congress of Virology,* Strasbourg, 1981, p 384.
3. Zuev VA, Pavlenko RG, Mirchink EP *et al:* Possible ways of modelling latent influenza infection in mice. *Vopr Virusol* 3:290–295, 1981.
4. Zuev VA, Mirchink EP, Kharitonova AM: Experimental slow infection in mice. *Vopr Virusol* 24:29, 1983.
5. Robinson JH, Easterday BC, Tumova B: Influence of environmental stress on avian influenza virus infection. *Avian Dis* 23:346–353, 1979.
6. Smolensky VI, Osidze NG, Bogautdinov ZF *et al:* Study of virus carrier in chicken influenza. *Vopr Virusol* 4:411–417, 1978.
7. Frolov AF, Shcherbinskaya AM, Rybalko SL *et al:* Phenomenon of prolonged circulation of influenza A virus in the body. *Mikrobiol Zh* 40:102–104, 1978.
8. Schild GC, Oxford JS, de Jong *et al:* Evidence for host-cell selection of influenza virus antigenic variants. *Nature* 303:706–709, 1983.
9. Patterson S, Oxford JS: Analysis of antigenic determinants internal and external proteins of influenza virus and identification of antigenic subpopulations of virions in recent field isolates using monoclonal antibodies and immunogold labelling. *Arch Virol* 88:189–202, 1986.
10. Katz JM, Maoliang W, Webster RG: Direct sequencing of the HA gene of influenza (H3N2) virus in original clinical samples reveals sequence identity with mammalian cell-grown virus. *J Virol* 64:1808–1811, 1990.

11. von Magnus P: Incomplete forms of influenza virus. *Adv Virus Res* 2:59–78, 1954.
12. De BK, Nayak DP: Defective interfering influenza virions and host cells: Establishment and maintenance of persistent influenza virus infection in MDBK and HeLa cells. *J Virol* 36:847–859, 1980.
13. Holland JJ, Grabeau EA, Jones CL *et al:* Evolution of multiple genome mutations during long-term persistent infection by vesicular stomatitis virus. *Cell* 16:495–504, 1979.
14. de St. Groth SF, Hannoun C: Selection of spontaneous antigenic mutants of influenza A virus (Hong Kong). *C R Acad Sci* (III) 276:1917–1920, 1973.
15. Golubev DB: Some actual aspects of antigenic influenza virus variability study. *Vopr Virusol* 1:117–121, 1975.
16. Golubev DB, Medvedeva MN: Experimental investigation of changes in the antigenic structure of influenza viruses during their persistence. *J Hyg Epidemiol Microbiol* 22:23–38, 1978.
17. Gavrilov VI, Asher DM, Vyalushkina SD *et al:* Persistent infection of a continuous line of pig kidney cells with a variant of WSN strain of AO virus (36405). *Proc Soc Exp Biol* (NY) 140:109–117, 1972.
18. Kantorovich-Prokudina EN, Semyanova NP, Berezina ON *et al:* Gradual changes of influenza virus during passage of undiluted materials. *J Gen Virol* 50:23–31, 1980.
19. Koch EM, Neubert WJ, Hofschneider PH: Lifelong persistence of paramyxovirus Sendai 6/94 in C129 mice: Detection of latent viral RNA by hybridization with a cloned genomic cDNA probe. *Virology* 136:78–88, 1984.
20. Stelmakh TA, Medvedeva MN, Golubev DB: Analysis of specific action between influenza virus and cells of different sensitivity: note 2. Characteristics of influenza virus–host cell interaction in persistent infection. *Rev Roum Med Virol* 33:47–51, 1982.
21. Monto AS, Maassab HF: Serologic responses to nonprevalent influenza A viruses during inter-epidemic periods. *Am J Epidemiol* 113:236–244, 1981.

15

Influenzal Anachronisms

THE EXISTENCE OF INFLUENZAL ANACHRONISMS

During the era of prevalence of a subtype, or of a major variant of H1 subtype, successive minor variants cause seasonal epidemics and then disappear. Occasionally evidence comes to light of the continued existence of the previous strain during the interepidemic period. Similarly, strains of a major serotype are sometimes found in the interpandemic period long after that serotype seemed to have vanished from the world population. More rarely still an outbreak of influenza cases may occur caused by such an anachronistic strain.

These anachronisms are providing information about the location and potential of influenza viruses during their supposed absence from the human community, and their significance varies according to the different sorts of anachronism.

THE PRESENCE OF THE PREVALENT STRAIN BETWEEN EPIDEMICS

If the prevalent strain of influenza virus could be frequently found during the intervals between successive epidemics, it would strengthen the current concept that it is surviving by continuous chains of direct transmissions. Such findings occur but they are infrequent.

In 1954, Zakstel'skaya[1] isolated A(H1N1 old style) strains from healthy donors at a blood donation session in Moscow at a time when no influenza was being reported in the Soviet Union.

During the 1950s, John Dingle and his colleagues in Cleveland, Ohio carried out a surveillance of families of Case Western Reserve University staff.[2] Throughout 1950 and 1951, more than 50 families numbering around 250 persons were visited each week by two field workers and a throat specimen collected. No influenza virus was isolated from these routine swabs from healthy persons. The search for carriers was intensified during the two influenza epidemics in 1950 and 1951 and the virus was once isolated from an asymptomatic individual.

Similar investigations in many parts of the world have shown that isolations of the current strain of influenza virus can be made between its epidemic seasons but that such isolations are infrequent. They indicate that the viruses are not altogether absent from the community but are too scanty to support the current concept that they are surviving by continuous transmissions.

The new concept proposes that persistent infection in carriers is produced by some such mechanisms as a balance between defective interfering particles and standard virions. The balance may at times tilt in favor of the standard virions, as happens in persistent infection of cell cultures in the laboratory, allowing temporary escapes of infectious virus.

THE REAPPEARANCE OF VANISHED VARIANTS OF THE PREVALENT SUBTYPE

Antigenic drift is not circular. It seldom reverses direction to reproduce past minor variants, therefore such a reappearance indicates that the strain has been somehow conserved.

When A(H3N2) subtype appeared in 1968, one of the earliest strains widely prevalent in Japan and other parts of the Far East was A/Aichi/2/68. Unlike its contemporary strain, A/Hong Kong/1/68, it disappeared in 1969. Ten years later in Adelaide, Australia, a 3-year-old boy developed severe croup and from his specimen A/Aichi/2/68 (H3N2) virus was isolated for the first time since 1969.[3] This 1979 strain was identical with the prototype of 1968 except that the nucleoprotein was distinguishable. There was no possibility that it was a laboratory contaminant.

The lad had not been born until seven years after the general disappearance of the Aichi strain. The virus would not have survived relatively unchanged for ten years by continuous transmissions in the human community without being detected. A conservative estimate suggests that such a process would have necessitated more than 1000 person-to-person consecutive transmissions. It is easier to believe that the Aichi viral genome had been stored in a 1968–69 sufferer, perhaps one of the boy's parents, and had been reactivated in the carrier in 1979.

ISOLATION OF STRAINS BELONGING TO A VANISHED SUBTYPE

All three human subtypes of influenza A virus have had cycles of prevalence separated by long absences. It is currently believed that the subtype may be conserved in some nonhuman species during their absences. The interpandemic isolation of strains of absent subtypes strengthens the concept that, on the contrary, their epidemiology is contained within the human species. Such isolations have been reported but they are rare.

In 1962, Klimov and Ghendon[4] isolated strains of A(H1N1) subtype six years after it had been replaced by the shift to A(H2N2) strains. The 1962 A(H1N1) strains differed from those of the 1946–56 era and from those of the later prevalence that began in 1977. Nevertheless, they show that the subtype was being conserved in the human community between successive eras of its prevalence. They were not reassortments between the old A(H1N1) and the then novel A(H2N2) subtype because they contained no genes of the prevalent A(H2N2) virus.

One of the earliest strains of A(H2N2) subtype known as A/Singapore/1/57 was isolated at Leningrad by Golubev and his colleagues[5] in 1980, 23 years after its disappearance and 12 years after the subtype had disappeared. They found the anachronistic strains in two outbreaks of influenza among Leningrad schoolchildren and in a contemporary sporadic case. The patients seroconverted against the H2 hemagglutinin antigen.

Anachronistic findings are sometimes attributed to faulty technique permitting escape of stored viruses in the laboratories. Golubev *et al.*[6] showed that the anachronistic strain differed from the A(H2N2) reference strains held in the laboratory.

OTHER SEROLOGICAL EVIDENCE OF ANACHRONISMS

In 1982, a date at which influenza A viruses belonging to A(H3N2) subtype were co-circulating with those of A(H1N1) subtype, Ivanova and her colleagues examined the sera of 652 children in Leningrad.[7] Some contained antibody to A(H2N2) subtype, which had disappeared 20 years earlier, long before the children were born. A few of the children were also found to possess antibodies to A(H0N1), a major serotype that had disappeared in 1946. Paired sera from 247 of these children showed some seroconversions to these past influenza viruses.

ANTICIPATORY FINDINGS BEFORE ANTIGENIC SHIFT

Change in the viral hemagglutinin has sometimes anticipated a change that subsequently occurred during antigenic shift. For example, the hemagglutinin of A/Dutch/56, one of the last minor variants of the A(H1N1 old style) era, was closer to that of the 1957 A(H2N2) strains than to that of earlier A(H1N1 old style) strains.[8]

Similarly, antibody to H3 antigen was gradually increasing in the population during 1966 to 1968, although the first A(H3N2) epidemic did not arrive until 1968.

In 1968, Hswine1 influenza A virus antibodies were found in children and antibody to A(Hsw1N1) strains was even higher in prevaccination sera in 1976, the year that Fort Dix suffered an epidemic caused by this virus.

In 1976–77, sera from unvaccinated children contained with increasing frequency antibody to A(H1N1 old style) strains before they reappeared in 1977. Monto and Maassab[9] commented that these antigenic variations must have been arising from within the human species, or perhaps in some cases from swine.

ANACHRONISTIC INFLUENZA EPIDEMICS IN REMOTE COMMUNITIES

The first documented anachronistic epidemic in which the causal influenza virus was identified occurred in Alaska in 1949[10] when the population was attacked by A0 virus, later known as A(H0N1), which had probably been prevalent from 1929 until replaced in 1946 by "A1" strains (AH1N1 old style). Even in 1949, the Alaskan communities were relatively remote from the main chains of human communication, living in a hostile environment between 60° and 70° N, about one third of the region lying within the Arctic Circle. The anachronism suggests that the Alaskans may have escaped the prevalent influenza during the A0 era, and that A0 strains persistently infecting human hosts had reached Alaska and been reactivated in 1949 among a susceptible community.

A serological study by Patricia Napiorkowski and Francis Black[11] among very isolated Amerindian communities of the Brazilian Amazon made a comparable discovery. The sera of several of the communities contained no evidence that anybody had ever been infected by the known influenza viruses. In one tribe, however, the Mekranoti consisting of 192 persons, all but two of the 61 sera collected from them in June 1972 contained antibody to A(H2N2) virus. The positive specimens came from persons of all ages including four children estimated from the dentition to be 16–25 months old. The two negative specimens also came from children. The Mekranoti must have been attacked by an epidemic of A(H2N2) influenza in the 1970–71 season, two years after the antigenic shift that replaced that virus by strains of A(H3N2) subtype.

Napiorkowski and Black make the following comment:

> We do not know whether the H2N2 virus had persisted in the Indian community because the other strains had not been introduced to displace it, or whether it had persisted at low frequency in the larger Brazilian community and had been introduced from there. Other evidence suggests that influenza would be unlikely to persist in a population as small as the Mekranoti (192 persons) for lack of an adequate supply of new susceptible persons.[11]

They are attempting to explain their findings in terms of the virus surviving solely by continuous direct spread from the sick. The anachronism seems more comprehensible if carriers of persistent infection or of the retained A(H2N2) genome had come into contact with this immunologically virgin community at a time when the latent infection in the carriers was being reactivated to infectiousness.

One must conclude that, taken by themselves, these various types of anachronism do not support the concept that the virus is surviving between epidemics and between eras of prevalence by continuous direct spread. They are not conclusive as to the part played by nonhuman hosts in the epidemiology of human influenza, but they favor the concept that during the last century human influenza has been an entirely human affair, with no need to invoke a role for alternative host species.

REFERENCES

1. Zakstel'skaya L Ya: in Zhadnov VM, Solov'yev V, Epshsteyn F (eds): *The Study of Influenza.* Washington, US Department of Health, Education and Welfare, Public Health Service, 1969, p 709.
2. Dingle JH, Badger GF, Jordan WS Jr: *Illness in the Home.* Cleveland, The Press of the Western Reserve University, 1964.
3. Moore BW, Webster RG, Bean WJ *et al:* Reappearance in 1979 of a 1968 Hong Kong-like influenza virus. *Virology* 109:219–222, 1981.
4. Klimov AI, Ghendon YZ: Genome analysis of H1N1 influenza virus strains isolated in the USSR during an epidemic in 1961–1962. *Arch Virol* 70:225–232, 1981.
5. Golubev DB, Galitarov SS, Polyakov YuM *et al:* Antigenic anachronisms of influenza viruses A(H2N2) in Leningrad in 1980. *Zh Mikrobiol Epidemiol Immunobiol* 11:56, 1984.
6. Golubev DB, Karpukhin GI, Galitarov SS *et al:* Type A influenza (H2N2) viruses isolated in Leningrad in 1980. *J Hyg* (Lond) 95:493–504, 1985.
7. Ivanova NA, Grinbaum, EB, Taros, LYu *et al:* Serological substantiation of continuing circulation of influenza A(H0N1) and A(H2N2) viruses in children. *Vopr Virusol* 27:667–671, 1982.
8. Gengqi L, Xinchang G, Chu CM (Zhu J) *et al:* Antigenic relationship between H1 and H2 of influenza A virus. *Sci Sinensis* 23:1061–1068, 1980.
9. Monto AS, Maassab HF: Serologic responses to nonprevalent influenza A viruses during intercyclic periods. *Am J Epidemiol* 113:236–244, 1981.
10. Fenner FJ, White DO: *Medical Virology.* New York, Academic Press, 1970, p 166.
11. Napiorkowski PA, Black FL: Influenza A in an isolated population in the Amazon. *Lancet* 2:1390–1391, 1974.

16

The New Concept in Detail

The purpose of this chapter is to draw together into a coherent epidemiology the various theses of the new concept set out in previous chapters. The series of propositions deals with the difficulties individually and each proposition is followed by a brief discussion.

It should be emphasized again that the word "new" in the title is not a claim of priority. Others have proposed theses that are incorporated in the theory. Insufficient attention perhaps has been paid in the literature of influenza to the seasonal behavior of the epidemics and of their causal viruses. A simple explanation of the seasonal character of the human disease seems also to provide explanations for most of the score of other influenzal phenomena that the current belief in direct spread is powerless to explain.

This is far from being a final statement. The propositions are given in detail so that they may be corroborated, corrected, or disproved. Kilbourne[1] has warned against plausible theories:

> Influenza—this much studied and least understood disease—is not only a disease but for some of us a way of life. Once challenged with its virus, the investigator, unlike the patient, is chronically stricken and is doomed to a life of servitude to its whims—and [to] endless debates with his colleagues—unless he is rescued by the early attainment of a high administrative position or seduced by the largesse of cancer virology. A neuropathic sequel of this affliction is a delusion in which influenza suddenly becomes comprehensible. The investigator then becomes totally inaccessible to human communication until the next pandemic occurs, after which he is either restored abruptly to sanity or is led away muttering something about a "new hypothesis." Perhaps for this reason the student of influenza is constantly looking back over his shoulder and asking "what happened?" in the hope that understanding of past events will alert him to the catastrophies of the futures [sic].

The new concept was advanced in 1979[2] with some epidemiological evidence and a list of 12 of the difficulties inexplicable by direct spread. Unknown to us, Kilbourne had already listed 15 such problems and the number now exceeds 20 (Chapter 18: The Problem List as a Totality). Further evidence for the new concept was published in a series of papers in the *Journal of Hygiene,*[3–6] and a review in *Epidemiology and Infection,*[7] the new title of that journal, examined the concept in relation to human and nonhuman hosts and in laboratory experience.

Since 1979, the theory has altered a little, and the rapid increase in our knowledge of influenza viruses has necessitated changes in the arrangement and details of the constituent propositions.

It has been difficult to find an appropriate name. An early suggestion, "The Latency and Seasonal Reactivation Concept of the Epidemic Process in Human Influenza," is descriptive but cumbersome, and the term *latency* would be including both true latency and the mode of persistent infection.

PROPOSITION 1: CONCERNING LACK OF SPREAD DURING HUMAN INFLUENZA

The influenza virus cannot usually spread from the sick person because it so rapidly adopts the mode of persistent infection in the epithelium of the respiratory tract. The human host becomes a symptomless noninfectious carrier after the illness.

Comment: Proposition 1 denies the current belief that the influenza virus survives by a continuous chain of transmissions from persons with influenza, though it allows that such transmission can occasionally occur. The proposition explains the low attack rate so often reported from outbreaks within households.

The mechanism of persistent infection suggested conforms to laboratory experience. Von Magnus[8] long ago found that heavy infection of chicken embryos with influenza virus induced the production of incomplete virions, "von Magnus particles," now called defective interfering particles (DIPs). They appear early in human infections and are known to interfere strongly with replication of standard infectious virions even when DIPs are present in very small numbers. They are used in the laboratory to initiate the persistent mode of infection in cell cultures of influenza virus.

One must place in the scales against this proposition the ease with which material from early cases of influenza can be made to infect chicken embryos, cell cultures of various sorts, some laboratory animals, and, less readily, nonimmune human volunteers. The proposition does not claim that the virus cannot be directly transmitted but proposes that the evidence precludes this as the mode of transmission in natural epidemic influenza.

PROPOSITION 2: CONCERNING PERSISTENT INFLUENZA VIRUS INFECTIONS

The influenza virus remains in the mode of persistent infection during convalescence of the human host and for many months thereafter. Its presence usually causes no further illness to the host, who is unable to infect his nonimmune companions unless the persistent virus is reactivated to produce standard infectious virions.

Comment: Proposition 2 explains the apparent absence of the influenza virus between successive epidemics. Persistent influenza virus infections have been maintained in laboratory cultures for more than one year. The persistent mode tends to be unstable at times because evolution of the virus is progressing within the infected cells.[9] The balance between the replication of DIPs, temperature-sensitive mutants, and standard virions may swing in favor of the infectious particles, but the swing is rapidly corrected.[10] Infectious virions may occasionally escape and stimulate secondary antibody increases, thus explaining the rare isolations of the virus in the absence of an epidemic and the rise in population antibody that sometimes precedes antigenic drift.

PROPOSITION 3: CONCERNING SEASONAL REACTIVATION OF PERSISTENT INFLUENZA VIRUS

The persistent influenza virus infecting human carriers is annually reactivated to infectiousness by a seasonally mediated stimulus. The carrier, usually without again falling ill, becomes highly infectious for a brief period and his nonimmune companions, if infected, rapidly develop an attack of influenza.

Comment: Proposition 3 explains why influenza epidemics are seasonal, why they are ubiquitous (the carriers being widely distributed in the world community), how they cease automatically when those infected by carriers have had their illness (by proposition 1, they cannot then transmit the virus and so they comprise the whole epidemic), and the reason that many influenza epidemics begin explosively in large areas with no connection between the earliest cases.

All seasonal phenomena are ultimately attributable to the variation in the solar radiation resulting from the 23.5° tilt of the plane of the daily rotation of the Earth relative to the plane of the Earth's annual circumsolar orbit. Influenza epidemics, being seasonal phenomena, must ultimately be caused by these variations in solar radiation. The seasonally mediated stimulus affects human carriers of persistent influenza virus wherever they are living in all parts of the globe, its timing depending broadly on the latitude of the locality. In the tropics it operates twice each year around the time of the equinoctial monsoons, whereas north and south of the tropics it operates in the colder months of the year. Proposition 3 therefore explains the "transequatorial swing" of epidemic influenza that occurs annually.

One scans the influenza literature in vain for an explanation of its seasonal epidemicity. Kilbourne[11] dismisses its importance when he points out that "the seasonal patterns of influenza are not invariant, and springtime epidemics of influenza have been noted." Nevertheless, he does not discount the influence and, with Schulman,[12] was among the first experimenters to show that transmission of the virus and its spread in mouse colonies held under controlled temperature and humidity were facilitated in winter. Shadrin[13] in the Soviet Union showed a similar effect of both season and latitude in humans infected with live influenza virus vaccine.

Proposition 3 predicates two well-separated influenza seasons in tropical communities around the spring and autumn equinoxes, and two less widely separate seasons, often overlapping, should be detectable by careful analysis of the influenza of temperate latitudes.

We are ignorant of the mechanisms whereby the seasonal stimulus is operating, indeed, they have scarcely yet been sought. We are similarly ignorant of the modus operandi of most other seasonal phenomena, but we are in no doubt that such mechanisms exist.

PROPOSITION 4: CONCERNING INFLUENZA EPIDEMICS "OUT OF SEASON"

Unseasonal epidemics of influenza may occur when carriers are rapidly transported from one hemisphere to the other when their colony of influenza virus has just been reactivated seasonally in the locality from which they have come. The unseasonable outbreak will not spread beyond their infected nonimmune companions.

Comment: Proposition 4 follows logically from proposition 3 and explains why unseasonable influenza epidemics such as that caused by A(H2N2) strains in Cheshire in June 1957 (Chapter 7: Why Do Influenza Epidemics Cease?) are usually brief and circumscribed. Careful investigation of such outbreaks would help to validate or discredit the new concept.

PROPOSITION 5: CONCERNING THE SPEED OF EPIDEMIC TRAVEL OF INFLUENZA

The rapidity with which influenza epidemics travel through the human population is determined by the annual movement of the seasonally mediated stimulus that reactivates the virus in ubiquitous human carriers and so provides the opportunity for epidemics to develop in its wake among their nonimmune companions. Since the speed of movement of the epidemics is thus dependent on an extraterrestrial rhythm, it is unaffected by the speed and complexity of human communications. Influenza epidemics must have traveled at the same speed in previous centuries as they do in the twentieth century because they reflect the inexorable movement of the reactivating stimulus and not the virus being transmitted directly from one case of influenza to the next case.

Comment: Proposition 5, the unanticipated consequence of propositions 1–3, provides a further method of testing the validity of the new concept. Chapter 17 describes investigations that seem to confirm this proposition.

Although human communications do not affect the speed with which epidemics travel through the world population, they have an important effect on the character of the epidemics through the relative distribution of carriers and nonimmune persons.

PROPOSITION 6: CONCERNING ANTIGENIC DRIFT OF INFLUENZA VIRUS

The carrier develops immunity against his infecting virus (the parent strain) long before he receives the seasonal reactivating stimulus. Thus when reactivation occurs, the virions reconstituted to be identical with the parent strain are commonly neutralized by his immunity. In every infection with influenza virus, mutations occur at a rate of about 1 : 100,000 replications, and usually these related minor mutants are also neutralized. However, a few carriers, especially children, produce antibody with a more limited repertoire that neutralizes the parent strain while allowing the minor variants to escape. These are the carriers who transmit influenza and cause the next epidemic among their nonimmune companions. The influenza will, in such cases, be caused by a drift variant of the parent strain.

Comment: Proposition 6 provides mechanisms for antigenic drift and its seasonal timing.

Chapter 9 ("Serious Difficulties in Explaining Antigenic Drift") called attention to a major difficulty in explaining in any hypothesis how influenza virus can undergo antigenic drift, because it had been shown that antibody raised against the parent strain neutralizes the drift mutants as effectively. The discovery by Natali, Oxford, and Schild[14] that some persons produce antibody that will allow the escape of the minor mutants has explained the transmission by the new concept, whereas there are still difficulties for the current belief in direct spread during the illness. The authors have shown that some children and a smaller proportion of young adults produce such antibodies with a limited repertoire, but the situation for older persons has yet to be investigated. Such persons with a narrow antibody response are proposed as the carriers who transmit infectious virions for the continued survival of human influenza virus, and they are always widely distributed throughout the world population.

The discovery makes no contribution to solving the difficulty faced by the current belief in direct spread. The mutant strains would be vastly outnumbered by strains identical with the parent strain if transmission were occurring during the illness before antibody had developed, and antigenic drift would be most unlikely to occur.

PROPOSITION 7: CONCERNING THE SEASONAL METAMORPHOSES OF INFLUENZA VIRUS

A strain of influenza virus that has infected numerous persons and caused influenza over a wide area in a particular season and has become a persistent infection in some persons will have produced a similar immune response in them and led to a similar assortment of mutants. At seasonal reactivation in the carriers the mutants will tend to escape the immunity of the carriers (as in proposition 6) and be transmitted. The mutants vary in evolutionary fitness and commonly a single strain is outstandingly fit and is selected by the nonimmune recipients. In this manner the parent strain automatically disappears from the whole area of its prevalence and is replaced in a single season throughout that area, whether large or small, by its successor(s).

Comment: Proposition 7 explains the so-called "vanishing trick," the disappearance in a single season of the strain that has been causing influenza over a large area in the previous season. It also explains the associated phenomenon, how it can be replaced throughout that whole area by a related minor variant in the season of its disappearance.

The previous prevalent strain has produced a similar immune response in those it has infected, as witnessed by the production of the antibodies against the surface proteins H and N whereby the infecting organism is commonly identified. It is reasonable to suppose that the same antigenic input and the similar immune response will precede the production of similar virus mutants.

The new concept suggests that donor and recipient both participate in the selective process. Chapter 9 ("The Behavior of Natural Killer Lymphocytes . . . ") describes a contribution made by cell-mediated immunity in the donor,[15] and the same chapter (see Table 9.1) gives examples in which selection of more than one mutant of good potential seems to have been made by the recipients.

This proposition explains the metamorphoses that are occurring seasonally at antigenic drift of minor mutants. The metamorphoses that involve change to a reassortant (antigenic shift) or to a major mutant are discussed in subsequent propositions.

PROPOSITION 8: CONCERNING ERAS OF PREVALENCE OF INFLUENZA A VIRUSES

In the human host species, three subtypes of influenza A virus have had eras of prevalence in which a family of related minor mutants has caused successive epidemics. A different H-coding gene distinguishes each subtype—H1, H2, and H3—and the N-coding gene of the H1 subtype (N1) differs from that in the other two subtypes (N2).

Within the H1N1 subtype three major mutants have themselves had eras of prevalence that resembled the eras of subtype prevalence. These major variants can be known as Hswine1N1-like, H0N1, and H1N1 old style strains. Each era lasts until virtually all nonimmune persons have been infected.

PROPOSITION 9: CONCERNING RECYCLING OF MAJOR VARIANTS OF INFLUENZA A VIRUS

All three subtypes have had more than one era of human prevalence separated by interpandemic periods lasting many years. The major variant of H1N1 subtype known as H1N1 old style is known to have had three eras of prevalence in this century. It is proposed that the reassortant strains H1N1, H2N2, and H3N2 and the major H1N1 mutant (H1N1 old style) were all developed at different times long ago and have been continuously recycled in the human host species, perhaps for centuries.

PROPOSITION 10: CONCERNING "INTERPANDEMIC" SURVIVAL OF INFLUENZA A VIRUS AND ITS REACTIVATION

When a person has been attacked by an influenza A virus for the first time in his life, not only does he become a carrier of persistent virus for a year or more, but he also may retain the genome of the virus in his respiratory epithelium for the rest of his life.

The genome is reactivated by the same seasonally mediated stimulus that operates antigenic drift (propositions 6 and 7), but it is not able to initiate a new era of prevalence until sufficient nonimmune persons have been born into the human community and a suitable window is open for it in the immunological status of the world population.

The reactivation of H2N2 influenza A virus in 1957 and of H3N2 strains in 1968 initiated subtype antigenic shifts the order of which had been determined by their evolutionary development possibly centuries before. The recycling of the H1N1 old style major mutant of H1N1 subtype in 1946 and 1977 was also predetermined but on a different time scale.

The eras of prevalence of the other major mutants of H1N1 subtype—Hswine1N1 from 1918 and H0N1 from 1929—may also have been predetermined recyclings, or alternatively, they may have occurred on those dates from sequential mutations within the H1N1 subtype causing major antigenic drifts.

The antigenic drift after initiation of a new era of a recycled strain follows an antigenic sequence that differs from that of its previous era.

Comment: Proposition 10 explains the location of the apparently absent virus between its successive eras of prevalence and how it may reappear relatively unchanged even after more than 50 years absence. It also explains the vanishing trick at antigenic shift. H2N2 strains had completed the immunization of virtually all the nonimmune persons in the community when these strains disappeared in 1968 with the worldwide appearance of H3N2 strains. With minor exceptions, H2N2 strains have not yet reappeared, but their "interpandemic" absence since 1968 does not yet approximate the duration of their earlier absence from 1900 until 1957. One should not, however, anticipate precise timings from a parasite as unpredictable as the influenza virus.

The H1N1 strains with their three major mutant variants seem to be following an independent line. The predecessor vanishes with the appearance of the successor perhaps because, like the reassortant strains, it has immunized all the available nonimmune persons, or perhaps in 1918 and 1929 because at those times a major mutation in the H-coding gene had occurred. The latter explanation would presuppose a drift of sufficient magnitude to render the world population susceptible to infection by the novel mutant. Either explanation could explain the vanishing trick and worldwide spread within a single season.

PROPOSITION 11: CONCERNING THE LINEAGES OF HUMAN INFLUENZA A VIRUS SUBTYPES

Two lineages of influenza A virus have had eras of prevalence in the human community for at least the last 100 years, namely, one containing the H1N1 subtype with its three major variants previously known as H0N1, Hswine1N1-like and H1N1 old style serotype. The major serotypes of H1N1 lineage seem to be unable to have contemporaneous eras of prevalence. Similarly, the two subtypes of the other lineage, H2N2 and H3N2, seem unable to have co-prevalent eras. During the last 100 years, strains of the two lineages have only twice been co-prevalent.

Comment: Proposition 11 draws attention to a situation that is well-known but remains unexplained. There have been long periods in the history of human influenza A during the last 100 years when strains of a single major serotype have caused all the reported isolates of influenza A virus. For example, from 1918 until 1929, Hswine1N1-like strains probably caused all the human influenza A; then from 1929 until 1946, H0N1 strains were on their own; and from 1946 until 1957, the H1N1 old style strains. H2N2 strains caused all the influenza A from 1957 to 1968 and thereafter H3N2 strains caused all the influenza A until 1977. It is likely that H2N2 strains had a solitary prevalence from 1889 to 1900 and were replaced by a solitary prevalence of H3N2 strains until 1907 (see Fig. 10.1).

The situation must be expressing important information about the nature of the major serotypes and about the means of their production and succession. All of them coexist frequently within influenza B virus, but the only co-prevalences of major influenza A virus variants have been the two contemporary eras of H1N1 old style strains with H3N2 strains from 1908 to 1918 and from 1977 to 1990.

Although the phenomena remain unexplained, it may be valuable to keep them clearly in view by including them among the propositions.

THE POSSIBLY UNIQUE EPIDEMIOLOGY OF INFLUENZA

The propositions, taken together, are attempting to present a coherent epidemiology of influenza, and attention should be drawn to the concatenation of very remarkable phenomena that characterizes the influenza virus parasitism of mankind. The minor antigenic variation known as drift is a feature possessed in common with numerous other microorganisms, and the major variations, mutant and reassortant, are also features possessed by other small parasites. What may be peculiar to influenzal epidemiology is the manner in which these antigenic variations are linked with other features. First, they all occur seasonally. Second, at antigenic drift, the previously prevalent virus usually disappears in a single season from the whole area of its prevalence. Third, the minor variant that succeeds it replaces it throughout that area in the season of its disappearance. Fourth, at major variation of influenza A virus whether the successor is a mutant or a reassortant, the family of strains that had been prevalent worldwide and may have been causing all the influenza A in the world community for a decade or more vanishes in a single season. Fifth, the major variant that succeeds it becomes established worldwide in the season of the disappearance of the predecessor. Sixth, the epidemics of influenza tend to be ubiquitous moving south and north across the globe annually. Thus the distribution of the major variants of the virus also tends to be ubiquitous, whereas several of its minor variants may be simultaneously present in the world community in the same and in different areas. When two major variants are co-prevalent, as in 1908–18, and 1977–90, they may both be ubiquitous in the world community.

To these phenomena one must add the recycling of major variants and their solitary prevalences for long periods followed by still longer absences between successive eras, and the fact that each era of prevalence of a major variant immunizes almost all the accessible nonimmune persons in the world community.

It may be premature to claim this behavior as unique among human parasites, but it must alert us to the possibility that we are facing an unorthodox or unfamiliar epidemiology and that we ought to consider even hypotheses that appear to contravene our previous experience of transmission of viruses.

In the next chapter we consider the manner in which the epidemiology of measles, often used in the construction of models of epidemicity, differs from that of influenza.

REFERENCES

1. Kilbourne ED: The molecular epidemiology of influenza. *J Infect Dis* 127:478–487, 1973.
2. Hope-Simpson RE: Epidemic mechanisms of type A influenza. *J Hyg* (Lond) 83:11–26, 1979.
3. Hope-Simpson RE: The role of season in the epidemiology of influenza. *J Hyg* (Lond) 86:35–47, 1981.
4. Hope-Simpson RE: Recognition of historic influenza epidemics from parish burial records: A test of prediction from a new hypothesis of influenza epidemiology. *J Hyg* (Lond) 91:293–308, 1983.
5. Hope-Simpson RE: Age and secular distribution of virus proven influenza patients in successive epidemics 1961–1976 in Cirencester: Epidemiological significance discussed. *J Hyg* (Lond) 92:303–336, 1984.
6. Hope-Simpson RE: The method of transmission of epidemic influenza: Further evidence from archival mortality data. *J Hyg* (Lond) 96:353–375, 1986.
7. Hope-Simpson RE, Golubev DB: A new concept of the epidemic process of influenza A virus. *Epidemiol Infect* 99:5–54, 1987.
8. von Magnus P: Propagation of the PR8 strain of influenza virus in chick embryos. II. The formation of "incomplete" virus following inoculation of large doses of seed virus. *Acta Pathol Microbiol Scand* 28:278–293, 1951.
9. Holland JJ, Grabeau EA, Jones CL *et al:* Evolution of multiple genome mutations during long-term persistent infection by vesicular stomatitis virus. *Cell* 16:495–504, 1979.
10. De BK, Nayak DP: Defective interfering influenza virions and host cells: Establishment and maintenance of persistent influenza virus infection in MDBK and HeLa cells. *J Virol* 36:847–859, 1980.
11. Kilbourne ED: *Influenza.* New York, Plenum, 1987, pp 268–269.
12. Schulman JL, Kilbourne ED: Seasonal variations in the transmission of influenza virus infection in mice, in Tromp SW, Weihe WH (eds): *Biometeorology II.* New York, Pergamon, 1967, pp 83–87.
13. Shadrin AS, Marinich IG, Taros LYu: Experimental and epidemiological estimation of seasonal and climato-geographical features of non-specific resistance of the organism to influenza. *J Hyg Epidemiol Microbiol Immunol* 21:155–161, 1977.
14. Natali A, Oxford JS, Schild GC: The frequency of naturally occurring antibody to influenza virus antigenic variants selected *in vitro* with monoclonal antibody. *J Hyg* (Lond) 87:185–190, 1981.
15. Trushinskaya GN, Zhdanov VM: The role of natural cytotoxic lymphyocytes (natural killers) in the pathogenesis of influenza. *Vopr Virusol* 1:103–110, 1988.

17

Some Tests of the New Concept

CONSEQUENCES

A consequence, if the new concept be correct, is that the world population must be regarded as always almost ubiquitously seeded with symptomless carriers of both influenza A and B viruses. The intensity with which the carriers are distributed must vary from one location to another as it must also vary from one season to another. The proportion of carriers in the community must increase progressively during the era of prevalence of each major serotype, but the effect of the increase would be offset by the accompanying increase in the proportion of immune persons diluting the accessible pool of susceptible subjects.

We should expect influenza to occur throughout the world population each season, sparsely in some seasons but more intensely in others. Even in seasons of sparse influenza the cases would be widely distributed throughout the global community, most cases occurring in the colder months in each hemisphere.

EPIDEMICS THAT OCCUR OUT OF SEASON

Even though most epidemics in Britain occur in the colder months, unseasonable outbreaks do occasionally occur. We mentioned (Chapter 7: Why Do Influenza Epidemics Cease?) an outbreak in May–June 1957 caused by the Asian A(H2N2) strain several months before that novel subtype caused its first epidemic in the general population in England.

The new concept predicts that an unseasonable outbreak would be likely to occur among nonimmune companions of a carrier whose quiescent virus colony had been reactivated during recent residence in the other hemisphere. Such an outbreak could not spread and become general because it would consist entirely of the persons infected from the reactivating carrier(s), the sufferers themselves being unable to transmit the virus during their illness.

Such unseasonable outbreaks provide a test of the validity of the new concept, and careful search should be made for symptomless potential carriers. A military establishment like that attacked in the summer of 1957 would be a favorable environment in which the event might be expected (see Chapter 8).

EVIDENCE FROM THE SPEED OF TRAVEL OF EPIDEMIC INFLUENZA

A surprising consequence of the new concept is the conclusion that epidemics of influenza must always travel through the world community at a constant speed, and that they have probably done so for the many centuries since the virus became well adapted as a human parasite. This conclusion follows logically from the hypothesis that the apparent traveling of an influenza epidemic is not reflecting movement of the virus, but the inexorable annual movement of the postulated seasonally mediated stimulus that recalls persistent virus to renewed infectiousness in ubiquitous carriers. Their infected nonimmune companions act as an indicator of this movement by developing influenza in its wake. The movement, being extraterrestrially determined, is unaffected by the speed of human communications except as that affects the distribution of carriers and their nonimmune companions.

In ancient times, when human communications were minimal and slow, when roads were few and the minor paths were mere tracks particularly difficult and dangerous in winter, influenza epidemics would have moved at the same speed as they do in the twentieth century. More small localities might have escaped the epidemic altogether in the ancient epidemics because of the poor communications, but the rapidity with which the epidemic traveled across the globe would have been governed, then as now, by the solar seasonal rhythm. Evidence of this invariant speed of travel would support the new concept.

EVIDENCE FROM THE ANNALS OF INFLUENZA

It is possible to obtain evidence of the speed with which influenza epidemics traveled in ancient times. The numerous reports of recognizable influenza epidemics by contemporary observers from the sixteenth century onward are mostly disappointing from this point of view, the authors being more interested in the application of various treatments than in recording the date on which persons fell ill. An early exception is the questionnaire by John Fothergill concerning the mild but widespread influenza of 1775. The questionnaire was circulated promptly when the epidemic ceased in the first days of December and most respondents answered rapidly although the results were not published until 1784.

The following observations are abstracted from Theophilus Thompson's *Annals of Influenza*[1] to which the page numbers refer:

1. *Fothergill, Dr. John. pp. 88–89. London. 6 December 1775.* The epidemic began about the beginning of November, and within a week became more general. From having been largely treated at home it now claimed the attention of the faculty of physicians and "for the space of near three weeks, kept them, for the most part, universally employed."

> If these physicians in the country, into whose hands this essay may come, will be so obliging as to mention the time when this epidemic made its appearance in their neighbourhood. . . . The united observations of the faculty at large, must greatly exceed the utmost efforts of any single individual, however warmly he may be disposed to promote the utility of his profession.

2. *Pringle, Sir John, Bt. pp. 89–90. London. Not dated.* He himself suffered the disease but gives no dates. His remarks about the epidemic in Italy, France, and the Netherlands suggest that it was contemporaneous there with the British epidemic.

3. *Heberden, Dr. William, Sr. pp. 90–91. London. 16 December 1775.* The epidemic began 28 October and lasted for three weeks.

4. *Baker, Sir George, Bt. pp. 91–92. London. January 1776.* Regarding the 1775 influenza he writes: " . . . many people both in this room and its neighbourhood, were attacked some days preceding 20th October." He too mentions that the same epidemic had attacked France, Holland, and Germany, and seemed to have been more lethal on the Continent.

5. *Reynolds, Dr. Henry Revell. p. 93. London. 29 January 1776.* The doctor's wife fell ill with influenza on 23 October, and all his children had it. On 2 November he visited several patients who had been suffering for several days.

6. *Cuming, Dr. William. pp. 94–96. Dorchester, Dorset. 25 December 1775.*

> . . . The epidemic disorder that has been of late so generally felt, not only over all this island, but in several parts of Europe: and probably its influence has been far more extenisive.
>
> From the middle of October several individuals complained of colds, which were considered as accidental, and but little attended to; but it was not, I think, till after the 10th of November that the malady became general.

He compares it with the epidemic of the winter of 1732 that had attacked most parts of Europe, America, and the West Indies, of which an account is given in volume 7 of Edinburgh Medical Essays.

7. *Glass, Dr. Thomas, pp. 96–102. Exeter. Not dated.* Colds and coughs had been more frequent than usual during the previous autumn,

> . . . But from the 8th of November the number of people who were continually coughing increased so fast, that it was soon evident that the epidemical colds, which began in London, as we were informed by the public papers more than a week before, had reached us.

The disease reached the Devon and Exeter Hospital on 11 or 12 November and attacked 173 people within a week, "being all the servants and patients then in the house, except two children; 162 of them were coughing together."

The city workhouse began to experience a similar outbreak within a day or two of that in the hospital:

> From Exeter the disease travelled towards Cornwall; about the 13th of November it arrived at Okehampton and Ashburton, and about the 15th at Plymouth. . . . by the 20th it had reached Truro; and before the end of the first week in December, had spread to all parts of that county [Cornwall].

8. *Ash, Dr. pp. 103–105. Birmingham. 2 December, 1775.*

> The epidemic, of which we had accounts in the public papers from London, made its appearance in this place about the middle of November; and no fresh subjects were attacked with it after the 7th or 8th December. The period of it did not exceed a month . . .

Dr. Ash's remark about 7 or 8 December indicates that his letter has been wrongly dated 2 December.

9. *White, Dr. W. pp. 105–107. York. 22 December, 1775.*

> This epidemic seems to have appeared rather earlier with us than in London: it was observed before the end of October, became general in the beginning of November, at which time many whole families were indisposed. I was myself seized with it on the 2nd of that month; and in a very short time, it became the most universal disease that hath been remembered with us. It was much abated by the first week in December, and seems now to have entirely left us.

He adds that his account may be entirely depended on for its exactness, being the result of his own observations conjoined with those of the faculty in York.

10. *Haygarth, Dr. John. pp. 108–111. Chester. Not dated.* "The epidemical catarrh of 1775 seized, in general, the inhabitants of Chester about the middle of November." It spread most universally from 15 to 25 November, and few people were attacked as late as 2 December. One case seen on 2 November began six days earlier, 27 October.

His correspondence with doctors in Wales brought useful and puzzling information. The disease had already become general in North Wales within three to five days of its general seizure of the Chester inhabitants, that is, 18–20 November. It began around 20 November in the remote Lleyn peninsula. In other parts of Wales it began nearly a month earlier, on 27 October.

11. *Pulteney, Dr. R. pp. 111–112. Blandford, Dorset. 18 December 1775.* "This disorder was earlier here than at London." Dr. Cuming's letter contains a reference to Dr. Pulteney: "From the middle of October (to which time Dr. Pulteney fixes the commencement of the disorder, when he himself was seized, though he was never confined by it). . . . "

12. *Thompson, Dr. William. p. 112. Worcester. 20 December 1775.* "This distemper became general here about the middle of November, and spread gradually to the country around."

13. *Skene, Dr. G. p. 112. Aberdeen, Scotland. Not dated.* "It began here near the end of November, and continued for four or five weeks; the second and third week it was very general"

14. *Campbell, Dr. D. p. 113. Lancaster. 18 February 1776.* Dr. Campbell was convinced that the disease is contagious. He says it prevailed in London for three weeks before reaching Lancaster, three days after Liverpool was being universally attacked. He says that Kirkby Lonsdale, 14 miles northeast, was attacked a week later followed, a few days after by Kirkby Steven.

Figure 17.1 illustrates the timing of the epidemic. Those who have attempted to study the timing of an influenza epidemic in a community will be aware of the difficulty of fixing the date of origin and termination. Nevertheless, the answers give a reasonably coherent picture of an epidemic of influenza not appreciably different from one that might have occurred 200 years later in 1975. The timing of the maximal impact is impressively contemporaneous in mid-November in most of the communities, and the differences are such as would be expected from a similar investigation at the present day.

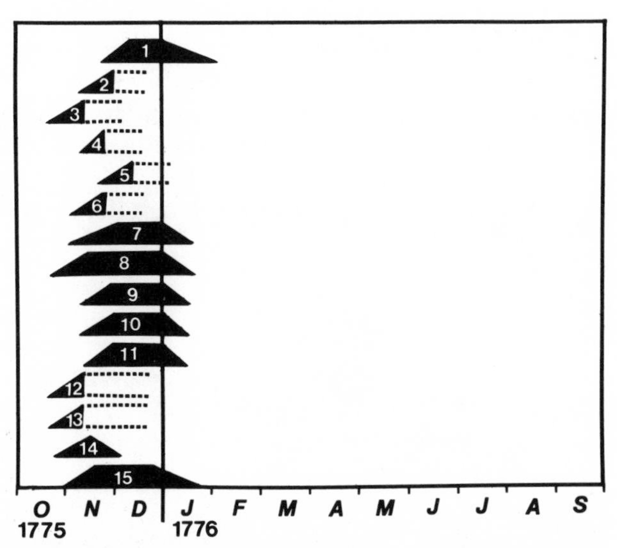

FIGURE 17.1. Dates of the geographical distribution of the 1775 influenza epidemic in the United Kingdom. Broken lines indicate that only the date of origin was recorded. 1. Aberdeen. 2. Worcester, 3. Blandford, 4. Welsh border, 5. Lleyn peninsular, 6. North Wales, 7. Chester, 8. York, 9. Birmingham, 10. Exeter, 11. Dorchester, Dorset, 12. London (1), 13. London (2), 14. London (3), 15. London (4) (data obtained from Thompson[1]; Hope-Simpson,[12] Fig. 1; reproduced with permission from *PHLS Microbiology Digest*).

EXCESS GENERAL MORTALITY ASSOCIATED WITH INFLUENZA EPIDEMICS

Some physicians in the eighteenth century paid particular attention to the mortality caused by epidemics of influenza. They noticed that not all influenzal outbreaks caused an increase in deaths over what would have been expected at the time of year, and indeed some epidemics were associated with a reduced mortality, as we shall see later.

The essay published by a Medical Society in Edinburgh describing the 1732–33 epidemic considered that the disease was not of itself fatal, " . . . but it swept away a great number of poor old consumptive people, and of those who were much wasted by other distempers."

These Edinburgh physicians were among the earliest to use the evidence in burial records (see Fig. 17.2):

> As a proof on whome it fell heaviest, we may remark, that, though the number of burials in Grayfriars churchyard (where all the dead of Edinburgh are buried) was double of what it used to be in the month of January, yet the number of those who were buried at public charge [the paupers] was so great that the fees of the burials scarce did amount to the sum commonly received in any other month. (p. 41)

Dr. John Arbuthnot also wrote an essay on the 1732–33 epidemic in which he noted that in London it lasted in its vigor from mid-January for about three weeks, and: " . . . the bill of mortality, from Tuesday the 23rd to Tuesday the 30th of January, contained in all 1588, being higher than any time since the plague [of 1665]" (pp. 36–37). Note how closely the elevation of the general mortality followed the heel of the influenza epidemic.

Huxham commented that the 1743 epidemic " . . . although exceedingly

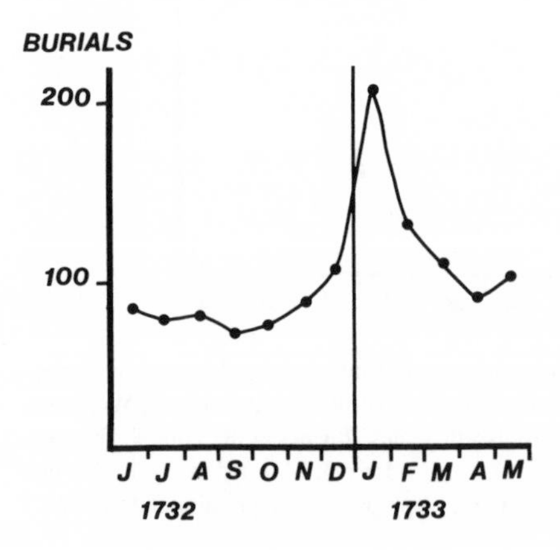

FIGURE 17.2. The 1732–33 influenza epidemic that peaked in Edinburgh in January 1733 was reflected by the excess of burials in Grayfriars churchyard (data from Thompson[1]; Hope-Simpson,[12] Fig. 2; reproduced with permission from *PHLS Microbiology Digest*).

common far and near, was fatal to few." Sir George Baker writes about the very widespread epidemic of 1762:

> In this city [of London], if the public records can be trusted, the burials during the prevalence of the disease did not much exceed the average. It is remarkable that at Manchester fewer than usual died when it prevailed. At Norwich, on the contrary . . . a much greater number fell victims than were destroyed by a similar pestilence in 1733, or by the more severe visitation, called Influenza, in 1743. (p. 76)

The 1775 epidemic was also reputed to have had a low mortality. Fothergill remarked:

> Perhaps there is scarcely an instance to be met with, of any epidemic disease in the city [London], where so many persons were seized, and in so short a time, and with so little comparative mortality. (p. 88)

Dr. Daniel Rainey was the physician in charge of the House of Industry in Dublin, an institution founded for the suppression of beggars and sturdy vagabonds, containing 367 paupers ranging in age from 12 to 90 years. More than 200 of them were attacked by the 1775 influenza, yet the governors of the institution reported that fewer had died during the epidemic than during any similar space of time since its foundation (p. 115).

The next considerable influenza epidemic, that of 1782, was investigated by the London College of Physicians, which reported that few had died except the aged, the asthmatic, and the debilitated (pp. 163–164). The London bills of mortality, however, to which they drew attention, tell a different story. There was a sharp, brief elevation in the number of burials associated with the epidemic period resembling that in Grayfriars, Edinburgh, in 1732–33 (see Fig. 17.3).

Dr. Falconer's account of the 1803 epidemic of influenza at Bath gives the number of burials in four parishes during the months preceding and including the epidemic, and concluded that "this disease was by no means so insignificant as it has been represented" (p. 271).

In 1837, the council of the Provincial Medical Association inquired into the influenza that had prevailed extensively during the first quarter of that year. The task of analyzing the replies to the detailed questionnaire fell to Dr. Robert J. N. Streeten. His attempt to determine the case fatality was not altogether successful, but the return from Dr. Black of Bolton in Lancashire clearly showed the excess mortality attributable to the epidemic compared with the average of the preceding five years (p. 301):

January 1837—excess mortality 3.8%
February 1837—excess mortality 126.0%
March 1837—excess mortality 2.2%

Dr. Shapter (p. 302) of Exeter makes a similar observation. The burials in the two

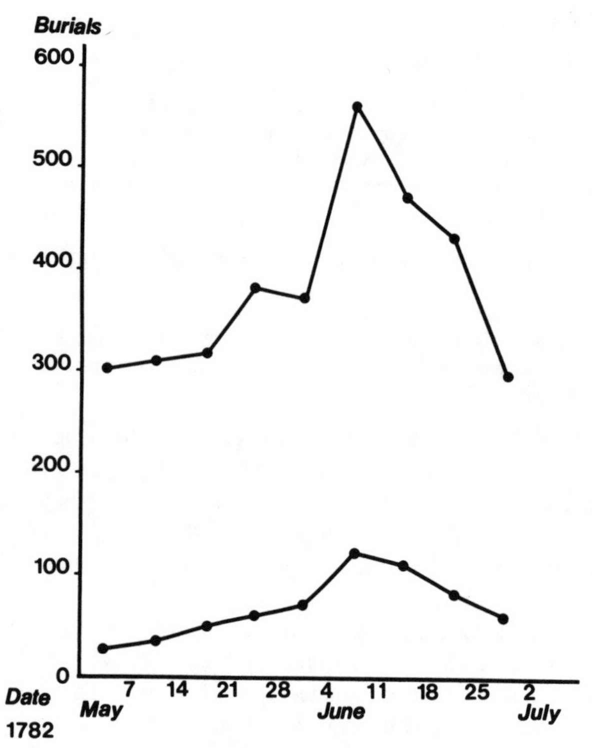

FIGURE 17.3. Sharp increase in excess general mortality in London in June 1782 attributed to influenza. The lower curve shows deaths attributed to "fever" (data from Thompson,[1] pp 163–164).

large cemeteries had numbered 125 in January and February 1836, whereas in the same months of 1837 they had totaled 227.

Perhaps the most interesting study from the mortality of 1837 comes from Dr. William Heberden, Jr., son of the Heberden quoted earlier in this chapter. Table 17.1, based on his table reproduced in Thompson's Annals (p. 340), compares the weekly burials with the christenings. In column 4 I have substituted the ratio of burials to christenings in place of his convention of relating burials to four christenings. The impact of influenza on the general (column 3) and specific (column 5) mortality is clear, and the age-specific figures emphasize the terminal role of influenza in the aged. It justifiably earned its name as "the old person's friend."

Dublin seems once again to have suffered the epidemic contemporaneously with London, as shown by Table 17.2, which records the burials in the Prospect cemetery that Dr. Graves (p. 341) considered to have buried rather less than a quarter of the persons dying in Dublin. It shows an indubitable excess of burials

TABLE 17.1. London Bills of Mortality and Christenings Showing the Impact of the 1837 Influenza Epidemic in January[a]

(i) 1837	(ii) Christened	(iii) Buried	(iv) Ratio	(v) Influenza	(vi) Age 30–40	(vii) 50–60	(viii) 70–80
January 3	363	228	0.63	0	14	20	22
10	487	284	0.59	0	23	42	30
17	384	477	1.24	13	49	70	53
24	520	871	1.67	106	69	95	122
31	307	860	2.80	99	71	54	113
February 7	532	589	1.11	63	41	69	77
14	474	558	1.18	35	54	70	59
21	316	350	1.11	20	36	36	31
28	809	321	0.40	8	32	24	37
March 7	480	262	0.59	4	23	23	19

[a]From Thompson,[1] p. 340.

in December 1836 to March 1837 compared with the same period in 1835–36, and the excess is by far the greatest in January 1837.

Dr. Theophilus Thompson himself witnessed this 1837 epidemic and he comments (p. 366) that the deaths in London were quadrupled during the prevalence of the disease. He also notes that the disease was prevalent in Berlin in the same month of January 1837.

The evidence available from the eighteenth and early nineteenth century shows that the epidemic was characterized by the same contemporaneity in widely separate localities as we find in the present century. Another feature that engaged the attention of the contemporary physicians was the sharp excess over the expected general mortality that accompanied or followed closely many though not all of the influenza epidemics.

TABLE 17.2. Burials in Glasnevin Prospect Cemetery, Dublin, Showing the Impact of Epidemic Influenza Especially in the Burials in January 1837[a]

	Month	Burials		Month	Burials
1835	December	352	1836	December	413
1836	January	392	1837	January	821
	February	362		February	537
	March	392		March	477

[a]From Thompson,[1] p. 341.

STUDY OF EXCESS OVERALL MORTALITY ASSOCIATED WITH INFLUENZA IN THE TWENTIETH CENTURY

A good discussion of excess general mortality attributable to influenza epidemics can be found in papers by Housworth and Langmuir[2] and Alling, Blackwelder, and Stuart-Harris.[3] The modern interest in the subject is often dated to William Farr's study of the impact in London of the influenza epidemic of 1847, though, as we have seen, much earlier observers had already studied mortality data. Like William Heberden, Jr., and others, Farr had called attention to the differential mortality rate in different age groups during influenza epidemics, the disease being most lethal for adults and especially the aged:

> The mortality in childhood was raised 83 per cent; in manhood, 104 per cent; in old age 247 per cent. From the age of 4 to 25, however, the mortality was comparatively not very much increased; at the age of 10 to 15, the healthiest period of life, it was scarcely increased at all in girls . . . [4]

Collins[5] and later Serfling[6] refined the methods for determining what rate of deaths should be considered as "expected" and "excess" at different times of the year. Their statistical techniques were principally concerned with attempts to forecast the visits and likely effects of influenza epidemics, and they have been used and elaborated by others such as Lila Elveback[7] in the United States, and Rvachev[8] in the Soviet Union.

Much earlier, Collins[9] had shown, as had been reported in previous centuries, that excess deaths during influenza epidemics were not confined to persons suffering from respiratory diseases, but were also to be found in groups suffering from nonrespiratory illness. Housworth and Langmuir concluded that computation of excess deaths attributable to influenza are best based on the mortality from all causes rather than solely on those attributed to influenza or to influenza plus all pneumonias as in some of Frost's studies.

EVIDENCE FROM PARISH BURIAL REGISTERS

There is thus overwhelming evidence that many influenza epidemics cause a sharp brief excess in general mortality. Is this a possible tool for exploring the problem that is so important for validating the new concept, namely, speed of travel of influenza epidemics in past centuries?

The observations quoted earlier in this chapter strongly suggest that such excess mortalities characterized many of the historic influenza epidemics, and the timing of them in relation to the epidemic differed little from that of the mortality from the disease in the twentieth century. The matter is of such importance that further evidence is desirable.

It seemed possible that burial registrations in British parishes, made compulsory by an act of King Henry VIII in 1538 and made more emphatically compulsory by a second act in 1558 during the reign of his daughter Queen Elizabeth I, might be used to fix the date of the presence of historic influenza epidemics in distinct localities.

The success of the technique depended on a number of factors, foremost being the existence of the characteristic excess mortality curves caused by epidemic influenza at a date when human communications were so much slower and less frequent that transmissions by direct spread would have been retarded. Some parishes might altogether have escaped the epidemic, and in ancient times the proportion of escaping parishes might have been much greater than at present. On the other hand, influenza attacking a tiny population such as characterized most rural parishes would tend to produce a mortality excess proportionately far greater than that in a larger population. The only way to discover if the technique was feasible was to attempt it.

In order to discover whether the excess mortality curve was recognizable, 13 parish registers in Gloucestershire were first studied. They were chosen alphabetically. The first hurdle was paleography. The earlier registers were written in a script in which neither letters nor numerals resembled current usage. Even when we had achieved fluency in reading them, some were so badly written or defective that they were useless. Most, however, were legible and some were beautiful.

Key years in which influenza epidemics had been recorded in the histories were chosen in four centuries, sixteenth to nineteenth. For the earliest dates epidemics were chosen that had been agreed to have been influenzal by medical historians. Only one such date was accepted in the sixteenth century. Some epidemics, as we have seen, were deemed by contemporary observers to occasion excess mortality while others, though widespread, were thought to have been nonlethal. Both sorts were studied.

The aim was to obtain a set of dates of the burials in each chosen parish for continuous 11-year periods that included each key year. Usually this was possible, but, when defective records precluded the complete set, a set of not less than five neighboring years was accepted for an average to compare with the key year.

The analyses for each study period compared the 52 weekly totals of burials in the key year with the average of the weekly totals of the 10 (or lesser number) of neighboring years. The graphs show the three-weekly totals, moving weekly.

Figure 17.4 illustrates one such chart from each of the four centuries. In each graph the key year shows a dramatic excess of burials in the parish coinciding closely with the date of the historic epidemic in the literature, although none of the historic observations related to Gloucestershire. For comparison, Figure 17.5 shows the excess of deaths registered weekly in Cirencester attributable to the 1918 influenza pandemic. Figure 17.6 illustrates two parishes in which epidemic influenza cannot be detected in the 1782 parish burials.

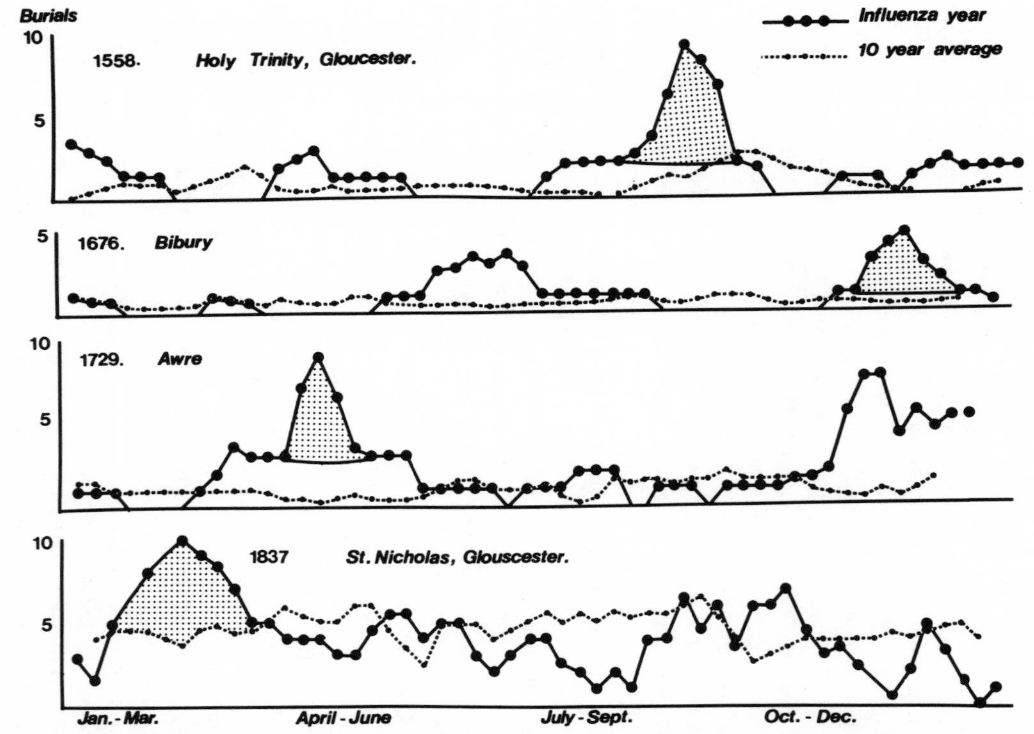

FIGURE 17.4. Examples of sharp excesses of burials in Gloucestershire parishes that coincided in four centuries with contemporary records of severe influenza epidemics. Three-weekly totals of burials moving weekly. (-●●●-, Influenza year; -•-•- , ten-year average.) (From Hope-Simpson,[10] Fig. 1; reproduced with permission from *Epidemiology and Infection.*)

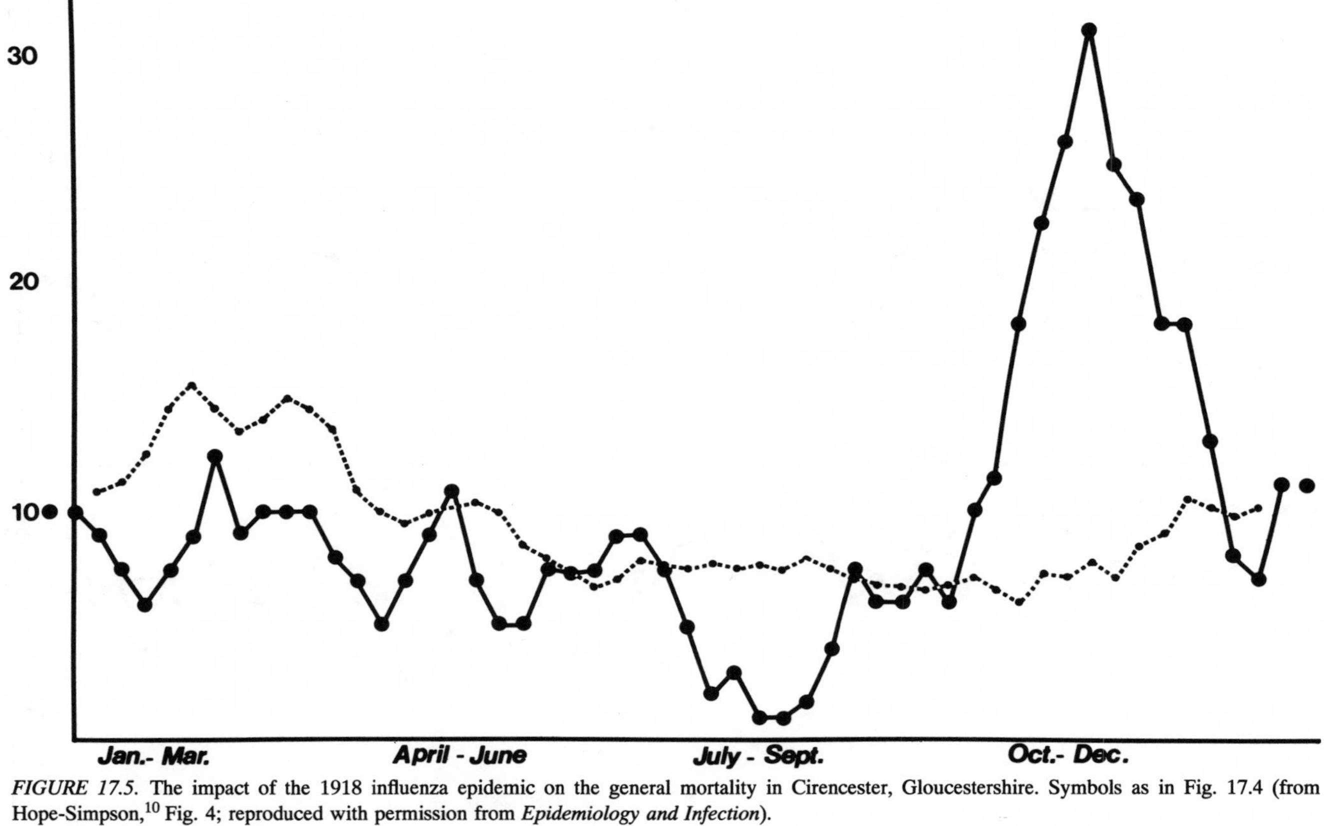

FIGURE 17.5. The impact of the 1918 influenza epidemic on the general mortality in Cirencester, Gloucestershire. Symbols as in Fig. 17.4 (from Hope-Simpson,[10] Fig. 4; reproduced with permission from *Epidemiology and Infection*).

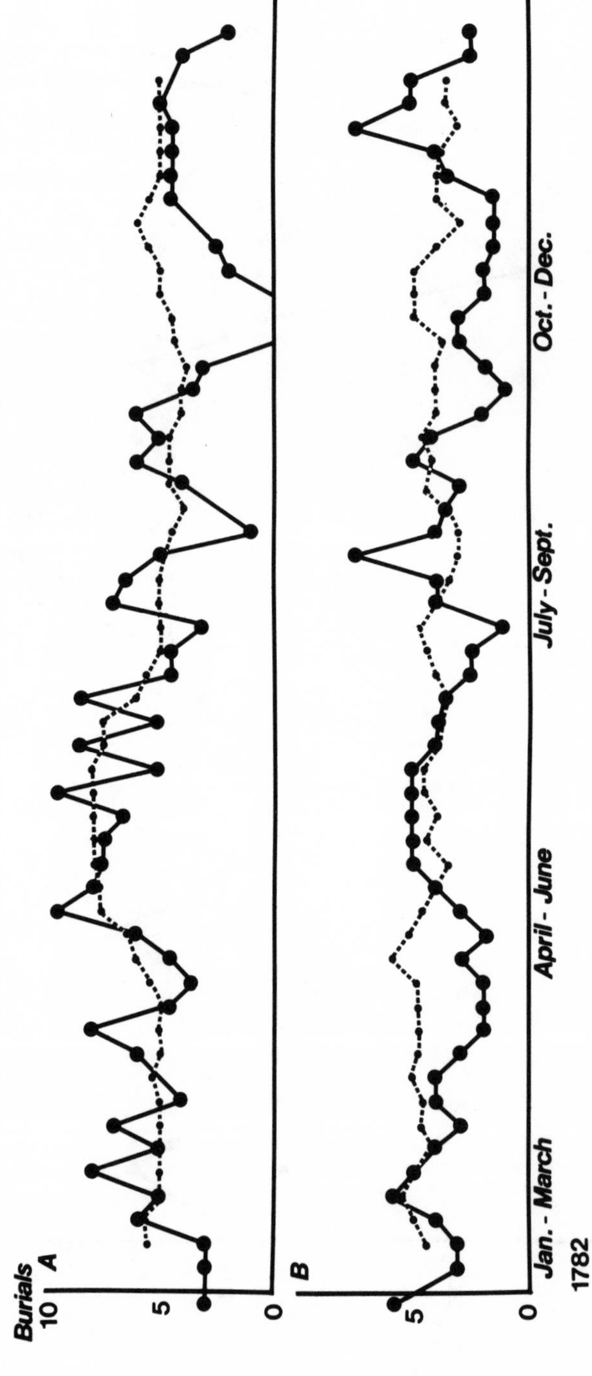

FIGURE 17.6. Examples from a year in which there was no noteworthy excess of burials (1782). (A) Burials registered in Cirencester parish. (B) Burials registered in St. Nicholas parish, Gloucester. Symbols as in Fig. 17.4 (from Hope-Simpson,[10] Fig. 5; reproduced with permission from *Epidemiology and Infection*).

It will be remembered that the 1775 epidemic studied by Fothergill and his friends was considered by some of them not to have been particularly fatal, and indeed it was remarked that fewer deaths than usual had occurred during the prevalence of the influenza. Figure 17.7, illustrating the burials in a Gloucestershire parish during 1775, shows the curve of burials actually dipping below the 10-year average in the timing of the epidemic in the contemporary writings. Evidently that small community was one of those that experienced fewer deaths than expected during those epidemic weeks.

During years in which severe influenza had been recorded, most of the parish registers studied showed the sharp elevation of burials well above the expected value at the appropriate time. It seems reasonable to accept the excess burials as evidence that the influenza epidemics, which we know were prevalent in England then, had at that time visited the parishes and caused the increased mortality. It may be objected that many other pestilences in olden days were causing increased

TABLE 17.3. Results of Examining Parish Records for Evidence of Excess Mortality[a]

(A) Twelve-month periods in which severe influenza is recorded and burial registers tended to show excess of burials around the time of epidemic prevalence (concordance).

Parishes in	1657–58	1728–29	1802–3	1836–37	1846–47	Total	% concordance
Cumbria	11/13[b]	11/14	6/6	—	—	28/33	81.8
Devonshire	5/6	6/7	8/9	5/7	—	21/29	82.8
Dyfed		5/5	5/5	5/5	4/5	19/20	95.0
Gloucestershire	3/6	9/11	10/11	3/5	4/5	29/38	76.2
Northumbria	3/4	10/17	14/16	—	—	27/37	73.8
Totals	22/29	41/54	43/47	13/17	8/10	127/157	
Percentage concordance	75.9	75.9	91.5	76.5	80.0	80.9	

(B) Similar periods in which an epidemic was reported not to have been lethal or in which there is doubt as to the influenzal nature of the epidemic.

Parishes in	1795–96	1831	Total	% concordance
Cumbria	5/12	—	5/12	41.7
Devonshire	4/8	3/7	7/15	46.7
Dyfed	0/5	2/5	2/10	20.0
Gloucestershire	1/9	1/5	2/11	18.2
Northumbria	5/16	—	5/16	31.2
Totals	15/50	6/17	21/67	—
Percentage concordance	30.0	35.3	31.3	—

[a]From Hope-Simpson.[11]
[b]Number of registers showing appropriate excess burials/total number of registers examined.

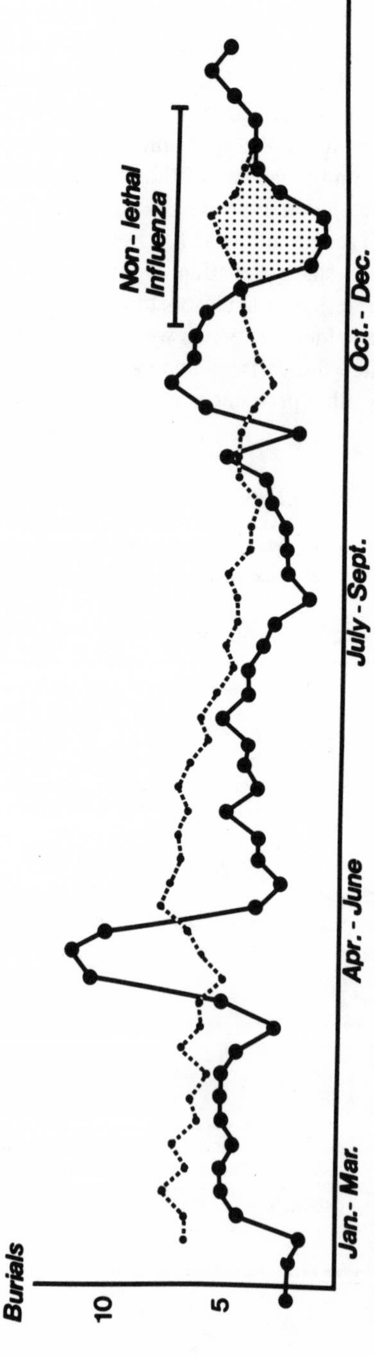

FIGURE 17.7. The decline in burials in Awre parish in autumn 1775 supports contemporary observations that fewer died than expected during that autumn epidemic. Symbols as in Figure 17.4 (from Hope-Simpson,[10] Fig. 3; reproduced with permission from *Epidemiology and Infection*).

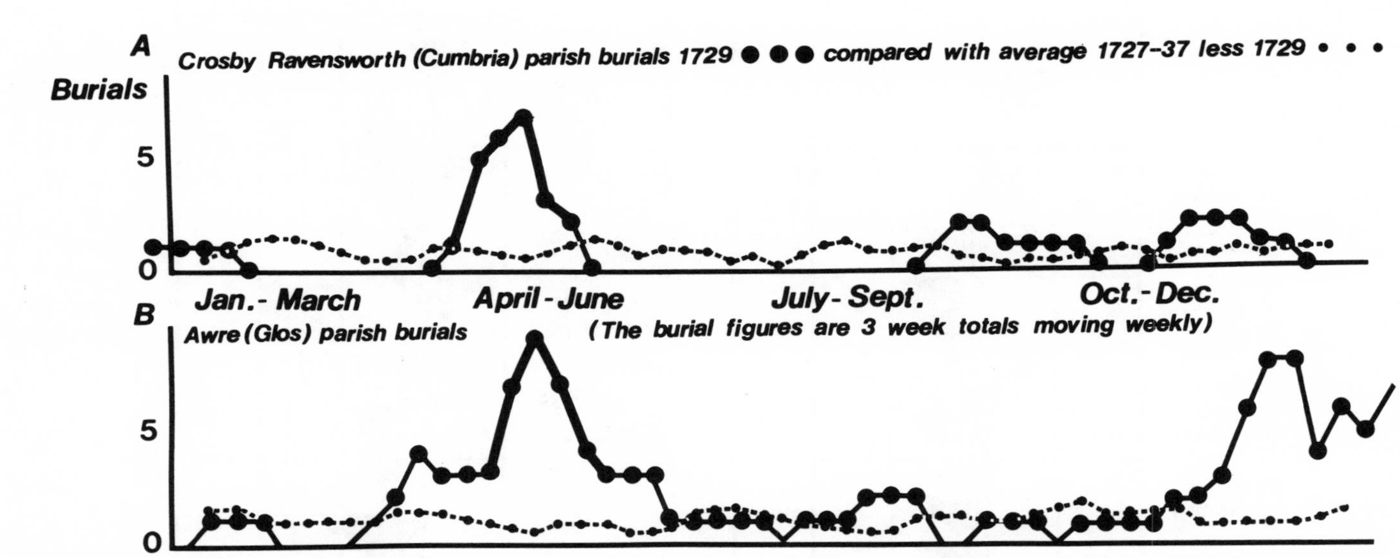

FIGURE 17.8. Synchronicity of excess burials. (A) Crosby Ravensworth, Cumbria. (B) Awre, Gloucestershire, 330 miles away, suggesting that the 1729 spring epidemic of influenza had attacked both rural communities contemporaneously. Symbols as in Fig. 17.4 (from Hope-Simpson,[11] Fig. 6; reproduced with permission from *Epidemiology and Infection*).

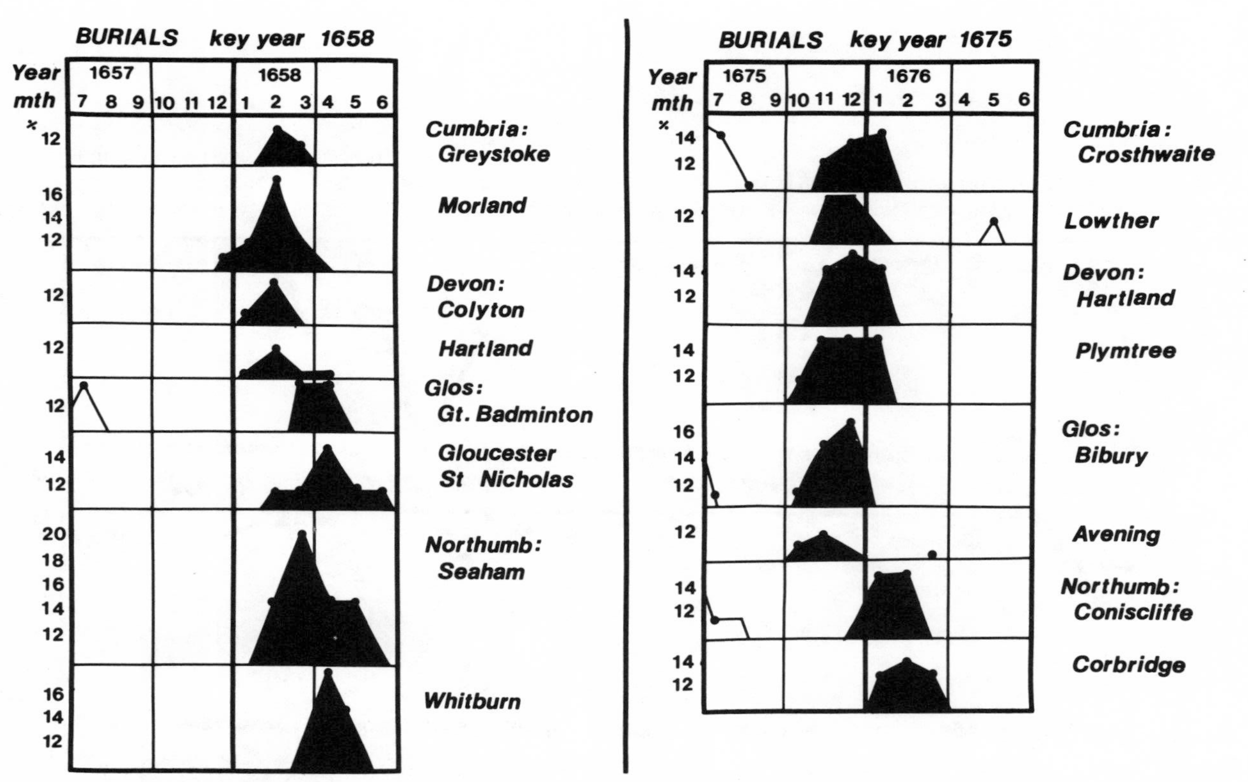

FIGURE 17.9. The timing of seventeenth-century influenza epidemics in diverse parishes in various UK counties as reflected in burials. The months are shown in which burials exceeded 10% of the 12-month total (from Hope-Simpson,[11] Fig. 1; reproduced with permission from *Epidemiology and Infection*).

mortalities, so that there can be no certainty that the phenomenon in these parishes was caused by epidemic influenza. This is true. In any particular case, one cannot claim with certainty that the date of the influenza epidemic did not coincide with that of some other lethal disease in the village community. However, the presence of the excess mortality in a majority of parish communities during the known prevalence of epidemic influenza is very strong evidence that the excess was caused by influenza.

The parishes that showed no excess at the expected time are of interest. They may have escaped the influenza epidemic, or they may have experienced it but had few or no deaths. In some such parishes we had evidence that a lethal illness had attacked the parish community in the recent past and culled the aged and infirm who would otherwise have succumbed to influenza, as in Figure 17.7.

This first part of the archive study had two consequences. First, it confirmed

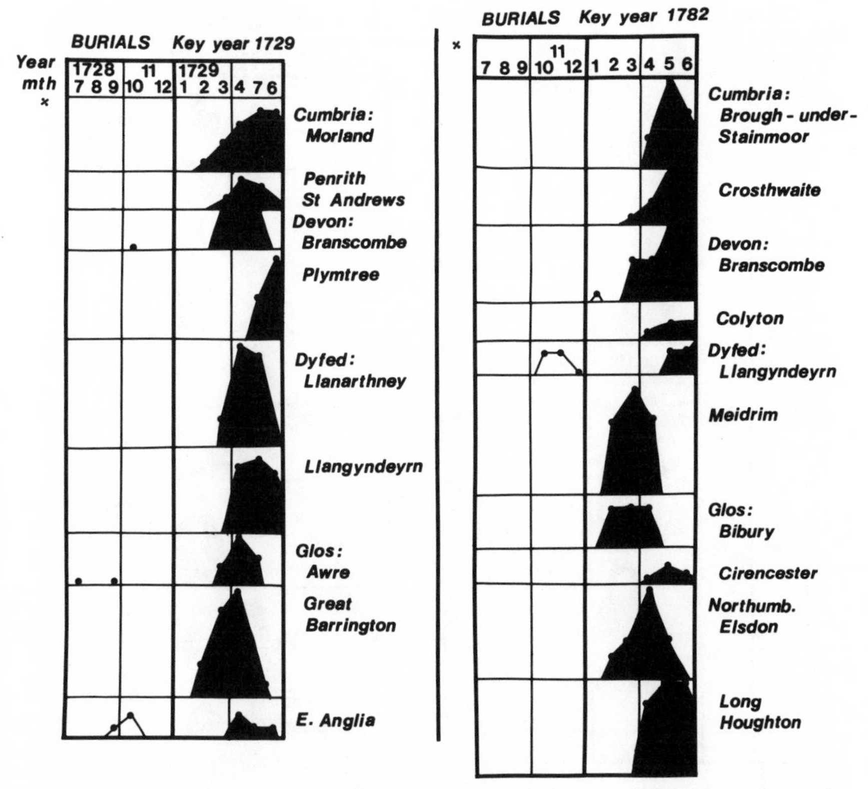

FIGURE 17.10. Excess burials in diverse parishes in England and Wales concordant with influenza epidemics in the eighteenth century. Symbols as in Fig. 17.9. See also Fig. 17.3 for London mortality (from Hope-Simpson,[11] Fig. 2; reproduced with permission from *Epidemiology and Infection*).

that the excess general mortality characteristic of severe influenza epidemics could be detected in these ancient burial registers, and second, that they occurred around the date anticipated from the historic records of influenza. Since such records had come from observers in counties other than Gloucestershire, the speed with which the Gloucestershire parishes were affected seems to support the hypothesis that influenza epidemics were traveling as rapidly in those past days as at present.

ARCHIVE DATA FROM WIDELY SEPARATED COMMUNITIES

The above conclusions could be further tested by applying the same technique to parish communities living in widely separated counties. For this purpose

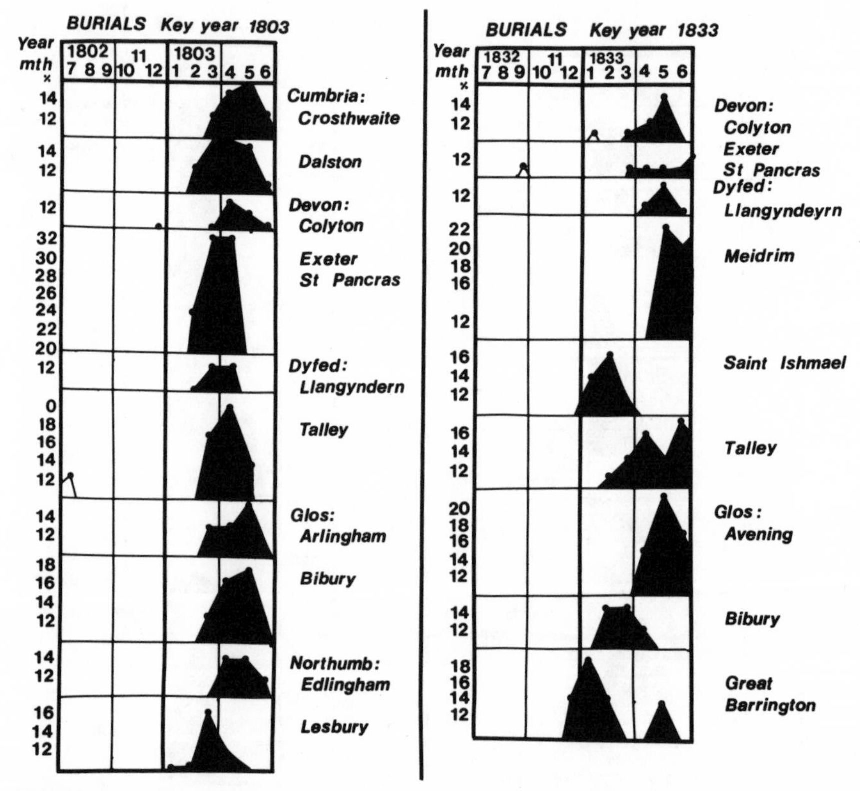

FIGURE 17.11. Excess burials in diverse parishes concordant with influenza epidemics in the nineteenth century. Symbols as in Fig. 17.9 (from Hope-Simpson,[11] Fig. 3; reproduced with permission from *Epidemiology and Infection*).

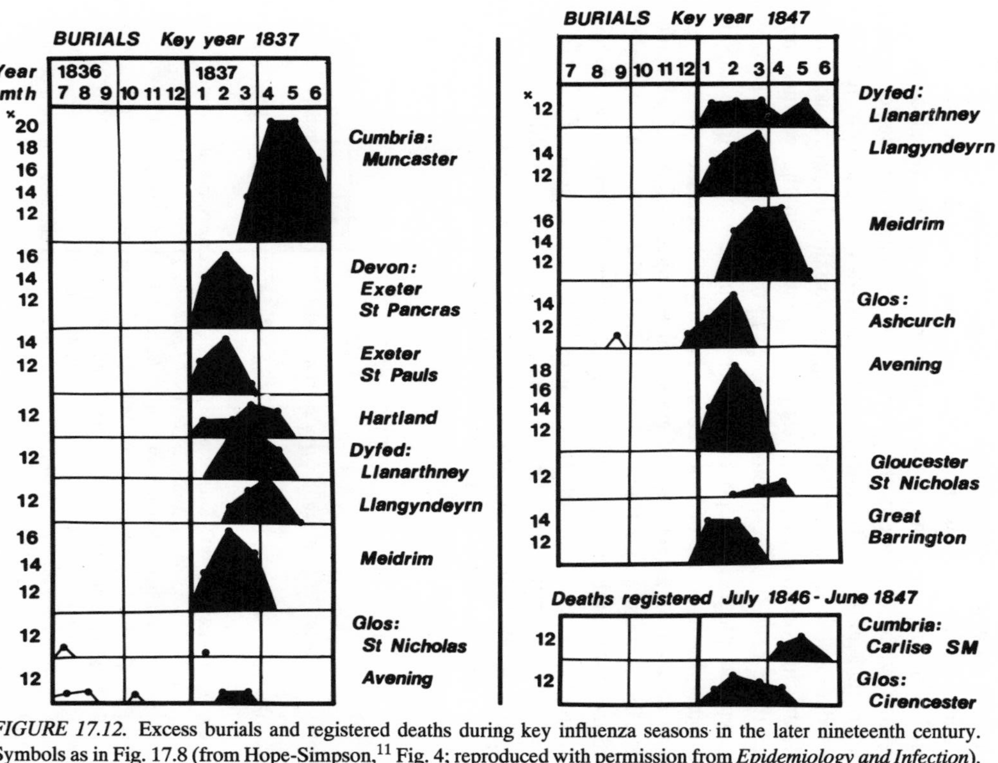

FIGURE 17.12. Excess burials and registered deaths during key influenza seasons in the later nineteenth century. Symbols as in Fig. 17.8 (from Hope-Simpson,[11] Fig. 4; reproduced with permission from *Epidemiology and Infection*).

burial registers were examined from parishes in Cumbria, Devon, Dyfed, and Northumbria and compared with those in Gloucestershire.

Table 17.3A shows the results of such analyses in five key years in which severe influenza epidemics had been recorded by contemporary observers. The concordance between excess mortality in the communities and the date of historic epidemics of influenza is too close to have been fortuitous.

Figure 17.8, which shows how closely the excess burials in a Cumbria parish coincided with those in a small community in Gloucestershire, also shows the absence of a winter increase in the average weekly deaths that is such a regular feature of nineteenth and twentieth century statistics. Surprisingly few of the parishes demonstrated this seasonal swing in mortality in previous centuries, possibly because deaths from nonrespiratory illnesses in the warmer months matched those caused by respiratory illnesses in the colder months. Dysentery, typhus, typhoid, plague, smallpox, and other ailments now uncommon must have caused many deaths in the summer during any 10-year period.

It is difficult to detect any difference in the speed of travel of epidemic influenza in the illustrations Figures 17.9 to 17.12. They show only the monthly proportions of the total annual burials that exceed 10%. The method is convenient but vulnerable to excess mortality from causes other than influenza. Earlier or later fatalities in the key year sometimes depressed the excess caused by influenza below the critical 10%. Nevertheless, the result is clear. Epidemic influenza appears to have traveled at an invariant speed during the last four centuries.

REFERENCES

1. Thompson T: *Annals of Influenza in Great Britain from 1510 to 1837.* London, The Sydenham Society, 1852.
2. Housworth J, Langmuir AD: Excess mortality from epidemic influenza, 1957–1966. *Am J Epidemiol* 100:40–48, 1974.
3. Alling DW, Blackwelder WC, Stuart-Harris CH: A study of excess mortality during influenza epidemics in the United States, 1968–1976. *Am J Epidemiol* 113:30–43, 1981.
4. Farr W: in *Vital Statistics.* London: Office of the Sanitary Institute, 1885, p. 330.
5. Collins SD, Lehmann J: Excess deaths from influenza and pneumonias and from important chronic diseases during epidemic periods, 1918–1951. Public Health Monographs No. 10 (PHS publications 213). Washington, DC: US Government Printing Office, 1953.
6. Serfling RE: Methods for current statistical analysis of excess pneumonia-influenza deaths. *Publ Health Rep* 78:494–506, 1963.
7. Elveback LR, Fox JP, Ackerman E *et al:* An influenza simulation model for immunization studies. *Am J Epidemiol* 103:152–165, 1976.
8. Rvachev LA: An experiment with computerized prognosis of influenza epidemics. *Dokl Akad Nauk SSSR* 198(i):68–70, 1971.
9. Collins SD: Excess mortality from causes other than influenza and pneumonia during influenza epidemics. *Pub Health Rep* 46:2159–2180, 1932.

10. Hope-Simpson RE: Recognition of historic influenza epidemics from parish burial records: A test of prediction from a new hypothesis of influenza epidemiology. *J Hyg* (Camb.) 91:293–308, 1983.
11. Hope-Simpson RE: The method of transmission of epidemic influenza: further evidence from archival mortality data. *J Hyg* (Camb.) 96:353–375, 1986.
12. Hope-Simpson RE: Simple lessons from research in general practice. Part 9. *PHLS Microbiol Dig* 8(1):23–26, 1991.

18

The Natural History of Human Influenza

INTRODUCTION

This book is calling attention to the many aspects of the behavior of epidemic influenza and of its causal viruses that the current belief appears woefully inadequate to explain, despite modifications that have been introduced to overcome difficulties. We have also described various alternative theories of which the new concept seems to provide the most comprehensive and plausible epidemiology.

Although it requires verification of some of its major hypotheses, for example, persistence of noninfectious virus after the influenzal illness, seasonal reactivation to infectiousness, and latency of the genome of the first infection with an influenza A virus, little discussion of the new concept has yet appeared in the literature since some of the propositions were advanced in 1979. The absence of comment should not be taken as an indication that students of influenza accept it. The opposite may be the case, namely, that the current belief in direct spread is so widely held and seems so self-evident that criticism of a contrary view is unnecessary. Those interested are content to await the findings of molecular virology, which seems poised to discover the existence of the virus in its persistent or latent mode if such exist after human influenzal infections.

Nevertheless, two reasoned criticisms of the new concept, both by persons who have themselves investigated the epidemiology of influenza, merit consideration in this chapter before our concluding attempt to describe the global natural history of the virus as a human parasite.

COMMENTS OF KILBOURNE ABOUT THE NEW CONCEPT

Kilbourne[1] discusses the new concept in his 1987 textbook on influenza as follows:

Hope-Simpson (1981) has proposed an all-encompassing role for season in the epidemiology of influenza, postulating "a direct effect of variation in some component of solar radiation on virus or human host." This hypothesis assumes latent carriage of the virus in human tissue, its seasonal reactivation in altered antigenic form, and its subsequent transmission to susceptibles to produce epidemics. This provocative hypothesis attempts to explain the apparent simultaneous eruption of epidemics caused by the same viral antigenic variant in widely scattered areas of the world. In the absence of any biological basis for influenza virus latency, and given the ubiquity of subclinical infections and the pervasive nature of human intercourse, current evidence favors the widespread seeding of epidemic viruses and their rapid propagation when ecological conditions are right. The seasonal patterns of influenza are not invariant, and springtime epidemics of influenza have been noted.

There is evidence for influenza virus latency in various modes in animal influenza and in laboratory culture. If subclinical infections are, as Kilbourne claims, ubiquitous, why should they be so and what is the host–parasite relationship? Are subclinical infections infectious for others and if so how is transmission brought about? What is meant by "widespread seeding of the epidemic viruses"? In what form are they seeded? How are they surviving until "ecological conditions are right"? Kilbourne continues:

But the idea of seasonal effects on host resistance cannot be discounted. In experimental infections of mice, transmission of virus was influenced by relative humidity but occurred more frequently in winter than in summer even when environmental controls provided identical conditions of temperature and humidity. In man, reactions to live virus vaccines have been reported to be more frequent in winter than in summer. The concurrence of influenza A and B emphasizes the importance of factors beyond the nature of the virus in the initiation of epidemics.

Any doubt about the communicability of influenza from those ill with the disease is dispelled by studies in crowded, confined or isolated populations.

He then gives instances that seem to him conclusive, although they appear not to have excluded the possible presence of symptomless carriers.

In 1963, he and Jerome Schulman had communicated their work on mice to a symposium at Pau, Béarn, France,[2] in which they said:

It is of interest, however, that when relative humidity was controlled, significant seasonal variations in transmitted infection were still present. . . . these seasonal changes [cannot] be explained by change in the virus, by differences in crowding or by stress due to exposure to cold since all experiments were exactly alike in design, differing only in the season when they were conducted. It is therefore necessary to find other explanations for the seasonal variations observed in these experiments. It is possible that such explanations may also be applicable to the seasonal variations in human influenza.

THE COMMENTS OF CLIFF, HAGGETT, AND ORD

Two geographers, Andrew Cliff and Peter Haggett, and a statistician, Keith Ord, have written a remarkable book called *Spatial Aspects of Influenza Epi-*

demics.[3] As Dr. David Tyrrell points out in his foreword, they have used powerful methods of computation and statistics in such a way as to reveal many points that remain obscure to those who simply handle the raw data. He urges all persons interested in influenza to read their book, a sentiment that I strongly endorse despite the criticisms that follow.

The main thrust of the book is the use of mathematical models both for elucidating the epidemic processes and, where possible, for predicting the consequences of epidemics. One chapter (pp. 45–86) is devoted to analysis of the Cirencester influenza data by time, microgeography, and antigenic nature of the viruses.

They give a brief résumé of the new concept in the course of a discussion of the transmission process in influenza (p. 18) as follows:

> The explosive simultaneous occurrence of influenza A in different locations, the relatively rare documentation of secondary cases within households, and the sudden appearance of influenza in isolated locations has led Hope-Simpson (1979) to propose a latent virus hypothesis. The hypothesis assumes that the influenza virus persists in some form in the human host and is reactivated, possibly in a genetically changed form, by seasonal triggers in a subsequent influenza winter. Under this scheme, an outbreak of influenza in one year would preseed the population to give a pattern of latent infectives from which a subsequent outbreak would arise in a later year. . . .
>
> In this book, we assume that most influenza cases recorded in a human population are generated by person-to-person transmission. We accept that cases may also occur by spread from animal or avian hosts and possibly from seasonally reactivated latent viruses.

On p. 72, in Section 3.3.5, entitled "Influenza as a Spread Process," they write:

> As we noted in section 2.2.3, Hope-Simpson has used the clinical records from his Cirencester practice to propose a wholly new theory of influenza transmission. The arguments for this have been fully set out in a series of papers . . . and will not be repeated here. Our own approach has been to subject the same data to geographical analysis and to interpret the results in conventional terms. We employ two major approaches—first, spatial autocorrelation analysis, and second, temporal autocorrelation and cross-correlation analysis—and consider each separately.

The book should be consulted for details of these methods and their results.

They conclude this section (pp. 77–78) as follows:

> One important aspect of Hope-Simpson's work has been his development of new hypotheses of influenzal transmission. In his (1979) paper, he reported that multiple case data from households within his practice showed little evidence of the four-day serial interval for influenza conventionally assumed to exist in most models of the spread of the virus between individuals. Hope-Simpson's analysis used data relating to spread of infection within households. . . . this absence of a serial interval led him to propose the latent virus hypothesis to account for the epidemiological features of the disease.
>
> At the much coarser geographical scale we have employed in this chapter, where Hope-Simpson's data have been examined in spatially aggregated form slightly different patterns emerge. At this scale influenza epidemics appear to peak sooner in urban Cirencester

> than in the surrounding rural area, and disease transmission from town to country is more rapid than from country to town. Within epidemics cycles have been detected in case levels at intervals of roughly three to four days and the harmonics thereof. These are the features we would expect if influenza spread can be regarded as a conventional person-to-person transmission process with a serial interval of three to four days.
>
> They may be attributed to the greater rate of person-to-person contact that occurs in towns compared with rural areas and to the fact that towns act as service centres for their rural hinterlands.

This explanation applies with equal force to the contacts between carriers and nonimmune persons. It is easy to overlook the fact that the new concept is also a concept of direct person-to-person spread, although the first of the two persons is usually a carrier, inconspicuous because he or she is symptomless. In their final paragraph on p. 78 the authors argue for the presence of a serial interval:

> Possibly the difference between Hope-Simpson's results and those described here are a function of different scales of analysis. We note that when Hope-Simpson's (1979) data are plotted, as shown in Figure 3.19 [reproduced here as Fig. 18.1B], there is some evidence, albeit very weak, of a slightly higher secondary case load at 4 days. In the absence of a serial interval, we would have expected [the figure] to show an exponential decay curve. It may be that the difficulty in establishing a well-defined serial interval for influenza arises because of a masking effect produced by the considerable variability in the lengths of transmission chains (between one and nine days). The issue remains unresolved and further work is necessary.

The reasoning here and in the careful analysis of the transmission process between individuals as set out in Section 2.2.3 (pp. 16–18) seems impeccable, based, as it is, on their general treatment of the data. They allow for all theoretically possible points of transmission in the infective chain proposed between donor and recipient. The serial interval might well have been masked except for the slight hint of it around day 4 (being their suggested average latent period plus the midpoint of the supposed average of the infectious period). Our own analysis of the Cirencester data (Fig. 18.1A), however, used household outbreaks in which transmissions from the introducing case would presumably have occurred early in the infectious period of the donor. Our analysis should therefore have revealed the presence of the serial interval more clearly than theirs had there been one to be disclosed. In fact it disclosed none in either of the first two A(H3N2) epidemics, and other studies have similarly failed to demonstrate its presence.

There is a further aspect of this part of the discussion that deserves emphasis. In order to support the direct spread hypothesis, the authors seem to be using estimates of latent and infectious periods already based on the assumption that the virus is being transmitted from case to case. The process has a dangerous circularity. To avoid the danger, we suggest that the question should be: "Which concept, direct case-to-case or symptomless carrier-to-case, is best supported by the distribution of the cases within affected household." We contend that the evidence in the data favors the new concept, and it is supported by the high proportion of such households, 70%, in which only a single case occurred both in

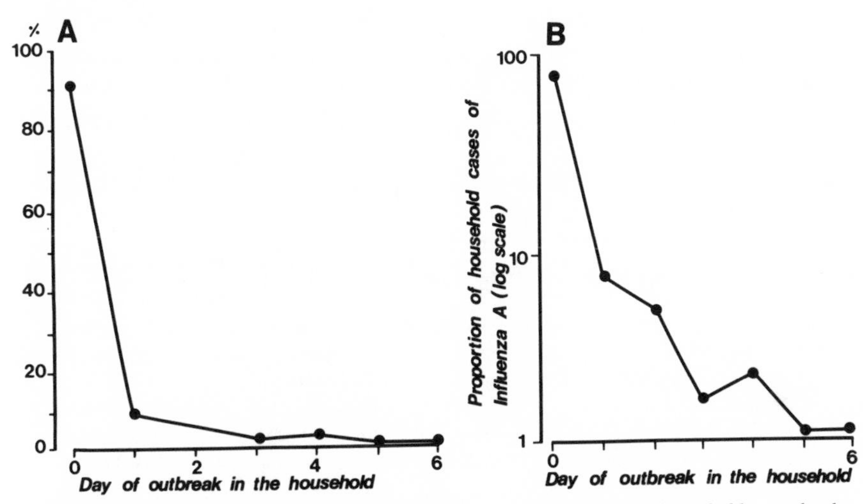

FIGURE 18.1. Two ways of looking at the distribution of the virus-positive household cases by day of household outbreak of influenza. (A) Daily percentage of the total. See also Figs. 7.5A and 7.6C. (B) The same data plotted on a log scale (from Cliff *et al,*[3] Fig. 3.19; reproduced with permission from Pion Limited).

the mild, protracted first A(H3N2) influenza epidemic early in 1969 and in its very severe successor eight months later (see Fig. 7.5).

ENDEMICITY AND THE CRITICAL POPULATION SIZE

Cliff *et al,*[3] in *Spatial Aspects of Influenza Epidemics,* have devoted most of their book to analyzing and explaining the temporal and spatial patterns of epidemic influenza. Iceland possesses good morbidity records, and they had previously been able to use them to analyze the behavior of measles in the inhabitants since 1896. The results were published in *Spatial Diffusion,*[4] that laid a sound mathematical foundation for work of this nature. They have used their experience to perform a similar great study of influenza in the same Icelandic population, and have therefore been able to make a comparison between the behavior of the two epidemic agents, measles virus and influenza virus, in that community.

Measles virus survives by direct spread from the sick patient to cause measles in infected nonimmune companions. It will die out in any community that is not large enough to recruit an adequate supply of susceptible persons by births during the time that it is passing around the community. For measles in civilized communities, it may be necessary to allow time for children to reach school age. In a population of sufficient size, measles will not die out, but will continue circulating in successive epidemics of varying size. Measles is said to be *endemic* in such a community. The epidemiologist, especially if he is interested in the mathematical modeling of epidemics, will need to know the endemic *threshold*, the minimum number of persons in the community needed to enable the particular agent that he is modeling to remain continuously in that community, without depending on reintroduction from outside the community. If, like measles, influenza depended on direct case-to-case spread for its survival in mankind, it would be important to know the endemic threshold, the critical population size that permits influenzal endemicity. The authors are interested to discover this threshold value:

> The survival of a particular virus to produce a continuous record of infection will be, among other things, a function of the population size of the community in which it is present. For endemicity to occur, sufficient individuals at risk must be present in the population at all times for the transmission chains . . . to remain intact. The critical population size has been examined by Bartlett [ref. 5] . . . for another communicable disease, measles. . . . Bartlett's analysis . . . indicated that a population of around 250,000 was necessary to ensure continuous transmission chains for measles. (p. 26)

Nine years later, Black[6] reported that an even larger community was required to sustain measles endemicity within island populations in the Atlantic.

Although there are host factors that affect the endemic threshold, for example, social and geographical dispersal, which may have raised the critical level for Black's island communities, the most constant factors concern the host–parasite relationship. Measles, with a case-to-case serial interval averaging around 11 days within the household and an infectiousness of around 75% among nonimmune household contacts, travels around a community exhausting the available susceptibles more rapidly than does mumps, which has a longer serial interval of around 18 days and a lower infectiousness of about 30–35%. The results are evident. Measles, even in a large family with numerous susceptible children, usually attacks all of them in a single generation of the disease after the introducing case, and the mother has them in bed at the same time. Not so mumps in such a family! The disease usually attacks one susceptible person after another so that the harassed mother of six nonimmune children wonders if she is ever going to get rid of the disease.[7] The threshold population that permits the endemicity of the mumps virus is much smaller than that needed to keep the measles virus in continuous circulation.

We have given particular attention to the endemic threshold because the matter closely concerns evidence of the correctness of the new concept of epidemic influenza as against the current concept of case-to-case spread, and it has been misunderstood by the authors of *Spatial Aspects of Influenza Epidemics.*[3] They say on pp. 26–27:

> Given that the serial interval for influenza is . . . approximately 2½ times less than that for measles, we might expect the endemicity threshold for influenza to be correspondingly reduced to a population size of around 100,000.

This is the opposite of the consequence to be expected from a shorter serial interval. Were the relationship as simple as the authors suggest, a figure between 700,000 and 1 million should have been mentioned. In fact, although it is true that a short serial interval and a higher infectiousness demand a larger endemicity threshold population, the relationship is more complex than that suggested in the quotation above.

The authors again attempt to justify the low endemic threshold from the Cirencester data in their Section 3.3.3 (pp. 65–67), though they remark that their figure (between 45,000 and 105,000) should be treated with extreme caution.

Their final discussion of endemicity, Section 6.2.1, is in relation to the epidemic influenza records from Iceland, comparing the population size with the number of months of reported cases of influenza (1945–70) in 50 "medical districts" of the island, and studying the results alongside those obtained for measles. They comment on their Figure 6.1 as follows:

> The significance of the two graphs is twofold. First, the parallel nature of the influenza and measles regression lines suggests that the size–endemicity ratio . . . for measles may also be extended to influenza. In this respect influenza appears to behave as an ordinary infectious disease rather than as having special mechanisms of spread. . . . Second, the difference in the endemic thresholds—a population of 290,000 for measles and 110,000 for influenza—is consistent with the difference in the length of the infection chains—typically twelve to fourteen days for measles and four to five days for influenza. . . . Thus, as far as influenza is concerned, even the capital city, Reykjavik, has been below the population threshold required for endemicity throughout this century. This helps to account for the repeating pattern of influenza epidemic waves . . . separated by quiescent months with no or few recorded cases.

Again the argument has been based on the false premise that the shorter the serial interval, the lower is the critical population for endemicity when the opposite is the case. Since such treatment of the data affects the construction of mathematical models of the disease the error is likely to have serious consequences for the models. The authors have graciously admitted the error in their calculations of the endemicity threshold.

HOW INFLUENZA DIFFERS FROM MANY OTHER EPIDEMIC INFECTIONS

Underlying many of the conceptual difficulties, there seems to lie a fundamental error. Measles and influenza differ from one another epidemiologically in an irreconcilable manner. Long-term studies of the epidemic behavior of measles, varicella, mumps, and infectious hepatitis had shown that there were three related features of the parasitism of such viruses with their human host species that were of great value in understanding the epidemic findings.[7,8]

The Serial Interval

This is an indirect measurement of the duration of the parasite's infective cycle with its human host. Each of these agents had its own characteristic serial interval, which is also the epidemiological evidence that direct case-to-case spread had been occurring.

The Infectiousness

This is determined by an estimate of the proportion of susceptible companions that catch the disease at each exposure in a particular environment such as the household. The infectiousness, which also had a characteristic value for each of the agents, cannot be determined accurately unless the serial interval is known.

The Age Distribution of the Persons Attacked

This depends in varying degrees on the urbanization of the community being studied, but in the same community the average age of patients attacked by each of these agents is related directly to the duration of the serial interval and inversely to the degree of infectiousness (Table 18.1).

Here is a powerful method for investigating the behavior of this group of immunizing infections and comparing it in different sorts of community. It seemed to have much to offer in the investigation of influenza A virus, which is also an immunizing infection within the era of prevalence of a particular subtype.

The optimism was misplaced. We have already told how in the careful household studies of the first epidemics caused by the Hong Kong A(H3N2) influenza virus no serial interval could be detected in cumulated household outbreaks, how the estimate of the attack rate subsequent to the first case in each household was only 17% in the first and only 14% in the much more severe second epidemic, and how the average age of those attacked was more than 30 years. It

TABLE 18.1. The Relationship between Serial Interval, Infectiousness, and Age of Susceptible Persons Attacked in the Cirencester Community[a]

Virus	Serial interval	Infectiousness	Mean age
Measles	10–11 days	75%	5½ yr
Varicella	14 days	60%	6½ yr
Mumps	18 days	30–35%	12 yr

[a]From Hope-Simpson.[7]

was ridiculous to suppose that the infectiousness of epidemic influenza is less than half that of mumps, and the absence of a serial interval compelled the thought that the epidemiology of influenza differs so radically from that of measles, varicella, mumps, and so forth that we needed to rethink the possible mechanisms and not attempt to interpret our findings by models based on measles (Chapter 7: Problems from Household Studies).

An example of the danger of conceptual error is the conclusion of Cliff and his co-authors[3] about their Figure 5.5 of influenza in Iceland (p. 139) (our Figure 18.2A):

> . . . as far as influenza is concerned, even the capital city, Reykjavík, has been below the population threshold required for endemicity throughout this century. This helps to account for the repeating pattern of influenza epidemic waves . . . separated by quiescent months with no or few recorded cases.

But the picture from the capital of Iceland is the same in essence as that from Houston, Texas and London, England, much larger communities, and as that from the tiny community of less than 4000 persons in Cirencester (Fig. 18.2). For hundreds of years, little and large communities all over the world have been experiencing such series of discrete influenza epidemics, and almost all of them, each on its own scale, has resembled the Icelandic picture.

THE PROBLEM LIST AS A TOTALITY

Apart from specific instances that we have been discussing in this chapter, neither Kilbourne nor Cliff has tackled the numerous difficulties that beset the current concept of case-to-case spread, and when the authors have defended it against a particular difficulty, the defense would often have been equally valid for the existence of silent carriers. For example, the priority claimed for epidemics in towns over their neighboring villages and the sometimes longer urban epidemics are also simply explained by the larger and more crowded population of the town and the fortuitous distribution of carriers and susceptible persons.

It is not enough to parry a few individual problems. Difficult as it is to achieve, Davenport's[10] dictum ought to be attempted. "Epidemiological hypotheses must provide satisfactory explanations for all the known findings—not just for a convenient subset of them." Concepts that fail to do so should give way to those that provide more satisfactory explanations, and they in turn may need to be superseded. Table 18.2 compares the adequacy of three concepts to explain most of these difficulties.

Thus considered, the current case-to-case spread concept appears totally inadequate even if the possible implication of alternative host species were to be accepted. The new concept fares better, but much has to be verified before it can

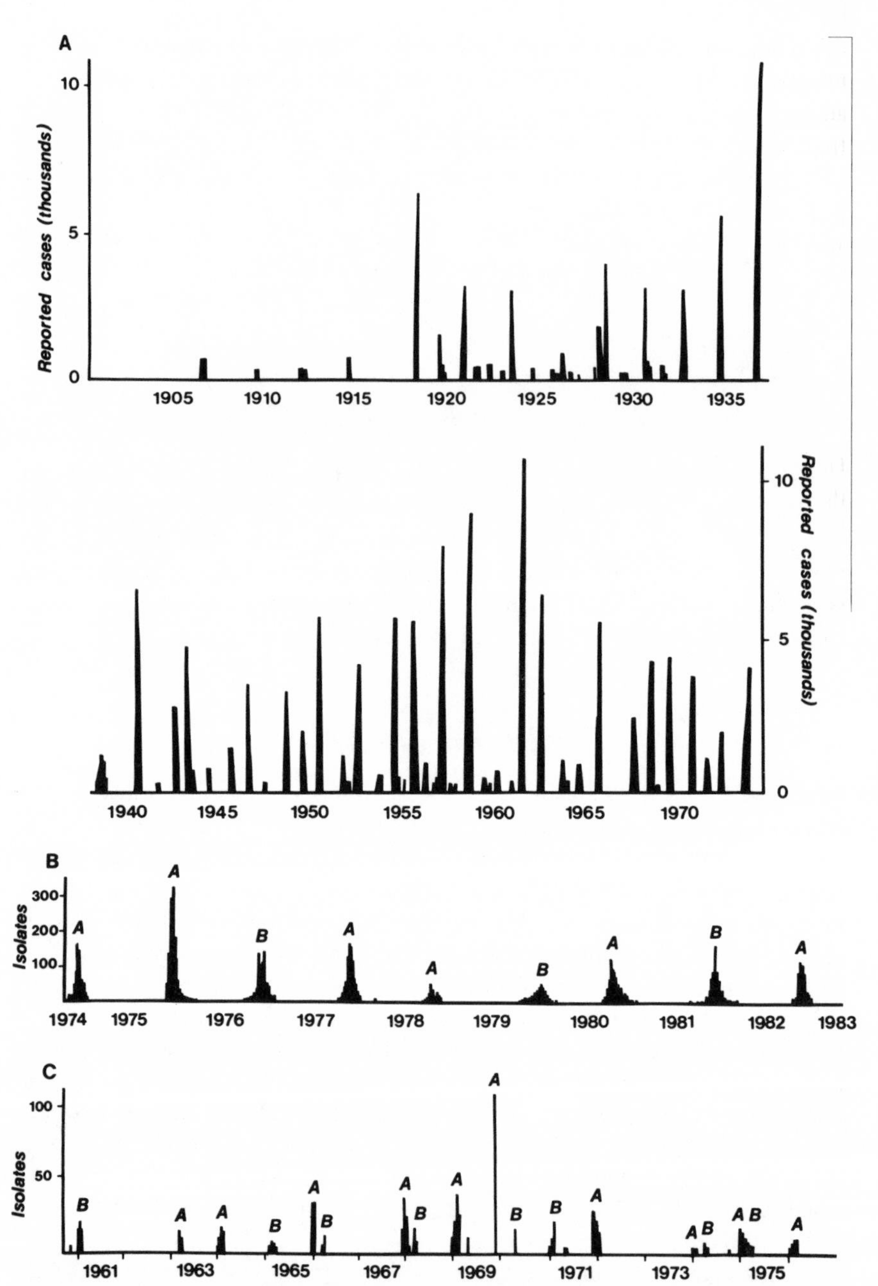

FIGURE 18.2. Secular distribution of influenza A and B epidemics in various populations. (A) Iceland (from Cliff *et al.*,[3] Fig. 5.5; reproduced with permission form Pion Limited). (B) Houston, Texas (from Glezen *et al.*,[12] Fig. 1; reproduced with permission from Academic Press–London). (C) Cirencester, England.

TABLE 18.2. Most of the Influenza Problems that Are Not Explained by Case-to-Case Spread (Current Concept), Even with Alternative Host Species, Are Comprehensible by the New Concept

Problem list	Current concept	Plus animal hosts	New concept
1. Ubiquity	—	—	+[a]
2. Seasonal epidemicity	—	—	+
3. Antigenic drift	—	—	+
4. Disappearance of prevalent strain	—	—	+
5. Prompt replacement over area	—	—	+
6. Interepidemic survival of virus	—	—	+
7. Epidemics explode over wide areas	—	—	+
8. Time and strains in small places similar to the whole country	—	—	+
9. Cessation of epidemics	—	—	+
10. Absent serial interval	—	—	+
11. Low secondary attack rate	—	—	+
12. Anomalous age distribution	—	—	+
13. Antigenic shift	—	?+	?+
14. Vanishing of major serotype	—	—	?+
15. Prompt replacement worldwide	—	—	?+
16. Recycling of major serotypes	—	?+	?+
17. Viral and serological anachronisms	—	?+	+
18. Seasonal antigenic changes	—	—	+
19. Out-of-season epidemics do not spread	—	—	+
20. Annual transequatorial swing	—	—	+
21. Constant speed, past and present	—	—	+

[a]Comprehensibility = +.

be accepted, particularly the demonstration that persistent infection is actually occurring in human influenza and continuing after the illness and that genome latency may follow it. From a theoretical viewpoint it is unnecessary to discover the mechanisms of the seasonal trigger. The fact of seasonal epidemicity is sufficient evidence that it exists, but understanding of the mechanism may be of critical value in designing prophylaxis against the disease.

IS INFLUENZA EPIDEMICITY UNIQUE?

During the first half of the present century, measles virus has survived in the United Kingdom in the manner shown in the diagram, Figure 18.3. It shows three lines of measles virus entering a community which, like most communities in this country, is largely immune to measles. Two of the chains of virus rapidly run into

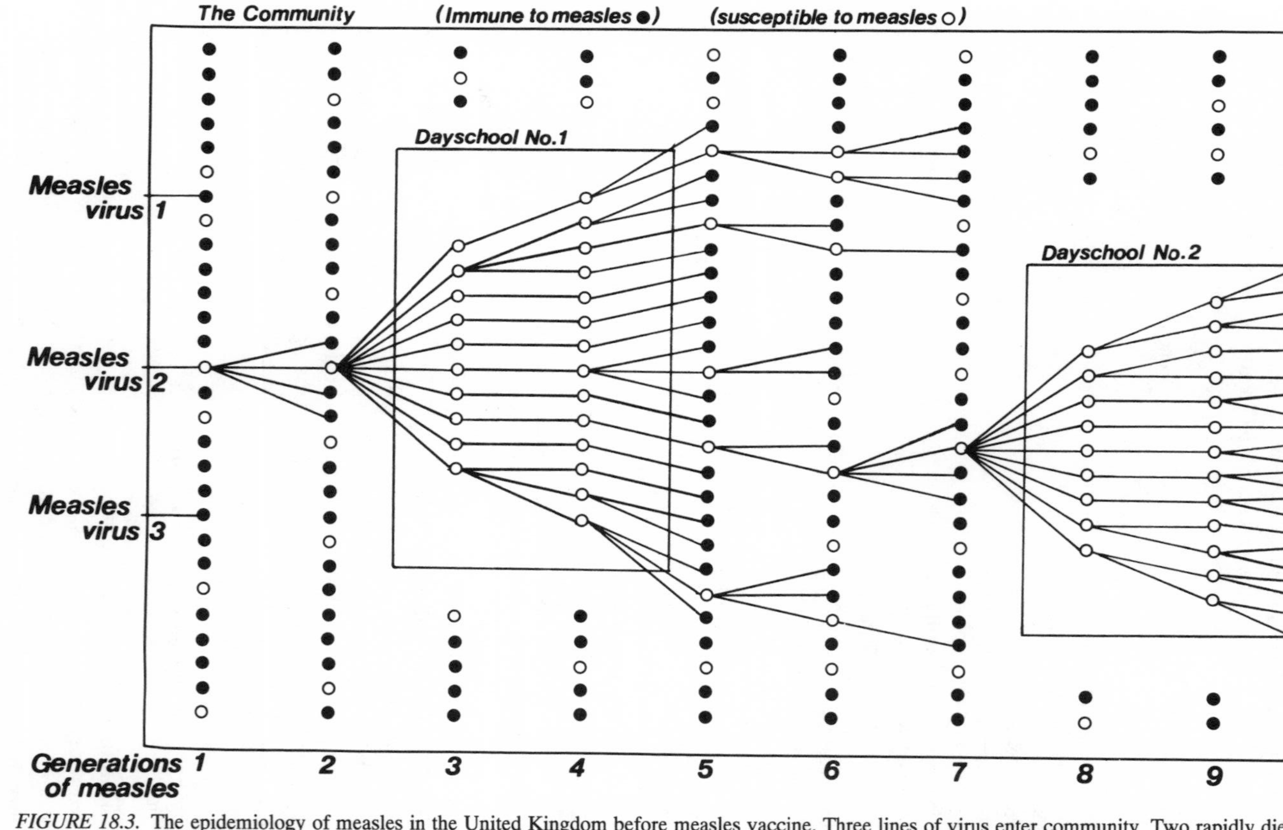

FIGURE 18.3. The epidemiology of measles in the United Kingdom before measles vaccine. Three lines of virus enter community. Two rapidly die out. One encounters a nonimmune child who infects the largely nonimmune day school. A boost of new lines spreads back into the community where most die out, but one line reaches another day school and repeats the boost (reproduced with permission from *Epidemiology and Infection*).

an immunological cul-de-sac and disappear, being unable to reach susceptible subjects for continued case-to-case transmission. The third chain happens on a susceptible day school pupil and is carried into the school in which most of the pupils have not before been exposed to measles virus. The school acts as a boosting station, numerous pupils coming in contact with the introducing child develop measles, and they in their turn transmit it to other pupils, so that several generations of the disease comprise the day school epidemic. The scholars carry the infection into their homes to infect their preschool siblings. In this way many new lines of measles virus again invade the general community, but most of them die out rapidly in the hostile environment. One line is fortunate enough to encounter a nonimmune child from another day school and that school also acts as a boosting station to promote measles virus survival. In large cities this process was continued repeatedly, and, by the time the circuit of all the day schools had been accomplished, sufficient new subjects had been born and attained school age to support the next round of measles in that city.

As Bartlett discovered, measles tends to die out in communities of under about 250,000. It did so in all the communities of smaller size, and they had to await reimportation of the virus from a larger community when their own population had recruited sufficient nonimmune children into their day schools to support another epidemic. Whereas the endemic–epidemic rhythm of measles in great cities had a two-yearly periodicity, in the Cirencester population (about 30,000 including the satellite villages) the measles epidemics came every four or five years, and in rural Wensleydale in Yorkshire, about every nine years during the first half of the twentieth century.

A similar picture on a different time scale characterized mumps epidemiology, but varicella was complicated by another source of the virus from patients suffering from shingles (*herpes zoster*), caused by reactivation of zoster-varicella virus that had remained latent in them since their attack of varicella.

Angulo and his colleagues[9] showed that smallpox virus was surviving by a similar epidemiological mechanism to that described above for measles. They adopted a geographical and mathematical approach, resembling that of Cliff and his colleagues, to study a smallpox epidemic occurring in a small Brazilian town. Their studies, which occupied the team for many years, are of especial value because smallpox has since been extinguished, one hopes, forever, and their careful work cannot be repeated. Their findings were summarized as follows:

> Computer-controlled contour-mapping of dates of introduction of variola minor into 169 households and the co-ordinates of the affected dwellings did not show a single contour pattern, but a group of areal patterns of within-household outbreaks. Introduction by adults and preschool children were distributed throughout the whole city area. However, introduction by school children formed two groups of contours and of affected dwellings. Each group was included in a discrete area corresponding to the zone of pupil recruitment of the two schools enrolling 91% of the school-child introductory-cases. The latter were responsible for introduc-

tion of the disease into 45% of the city's affected households. Altogether, both zones practically covered the whole city area. In either zone, several patterns surrounded the corresponding school. . . . contour maps clearly evidenced the influence of those two schools on spread of the epidemic.

The pattern of influenza is quite different. The day schools in the great cities are attacked as part of the general epidemic in the community, and they seem not to act as boosting stations from which the community derives its infection. The very severe epidemics of 1957 (H2N2) influenza A virus and 1969–70 (H3N2) lasted only six weeks and five and a half weeks, respectively, in Cirencester.

The first cases in the households examined during the 1951 epidemic of influenza were seldom school children. The contrast with measles is well shown in Table 18.3. One cannot doubt the importance of the measles-infected schoolchild in bringing the infection home, whereas the schoolchild has no such importance in the spread of influenza.

How, then, can we visualize the epidemicity of influenza? If influenzal infection caused an illumination, the northern hemisphere above the tropics would be ringed with lights at night during each northern winter sometime between October and March. As these lights died out they would be replaced by a belt of lights in the tropics in March and April. Then, as north tropical and south tropical lights faded, the southern hemisphere south of 23.5°S would become similarly decorated at night between May and August. Subsequently, as they began to fade, the tropics would again be lit in September. As the north tropics faded, the more northern communities would light up again to repeat the process for the next year. And so it would continue year after year, some years and some areas putting on

TABLE 18.3. The Percentage Distribution of Primary (Introducing) Cases and Secondary Cases in Household Outbreaks of Measles Compared with Influenza (Day 0 and Subsequent Household Cases) in Cirencester in 1951[a]

Age of patient		Measles	Influenza
Pre-school	Primaries	20	32
	Secondaries	80	68
	Total	100	100
School age	Primaries	80	32
	Secondaries	20	68
	Total	100	100
Post-school	Primaries	0	38
	Secondaries	100	62
	Total	100	100

[a]Adapted from Hope-Simpson,[11] Table II.

a splendid show, in other seasons scarcely worth watching (Fig. 18.4). For comparison, Figure 18.5 shows the same epidemic by longitude instead of latitude.

Thus the epidemics of influenza are swinging to and fro, south and north across the globe year after year. If any agent other than the influenza virus behaves in the same way, it has yet to be described. Measles, varicella, mumps, pertussis, rubella, infectious hepatitis, and smallpox do not do so. It is too early to know if influenza virus epizootiology behaves similarly in any host species other than mankind.

THE WIDER IMPLICATIONS OF THE NEW CONCEPT

As is the way of such things, the evidence supporting the new concept was obtained over a decade or more, and it was published in half a dozen consecutive papers. Great interest was expressed at seminars and lectures, and the concept was debated in the Soviet Union and the United States. The writing of this book was undertaken because it was pointed out to the author that many persons who do not see these specialized journals would be fascinated to read about the epidemiolog-

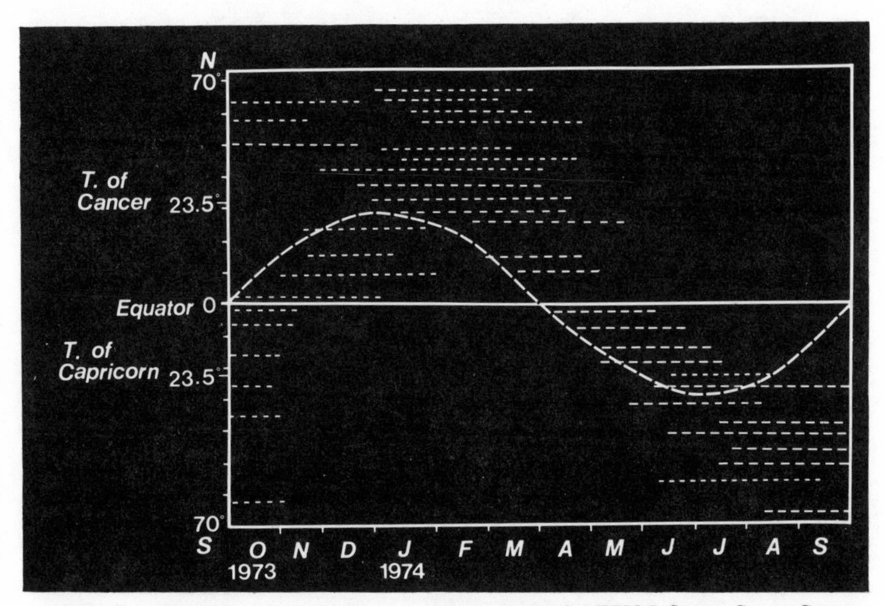

FIGURE 18.4. The timing of global influenza epidemics reported to WHO Influenza Center, Geneva, October 1973 to September 1974, shown by latitude (broken lines). The curve shows the annual shift of midwinter around the globe (from Hope-Simpson,[14] Fig. 1; reproduced with permission from *PHLS Microbiology Digest*).

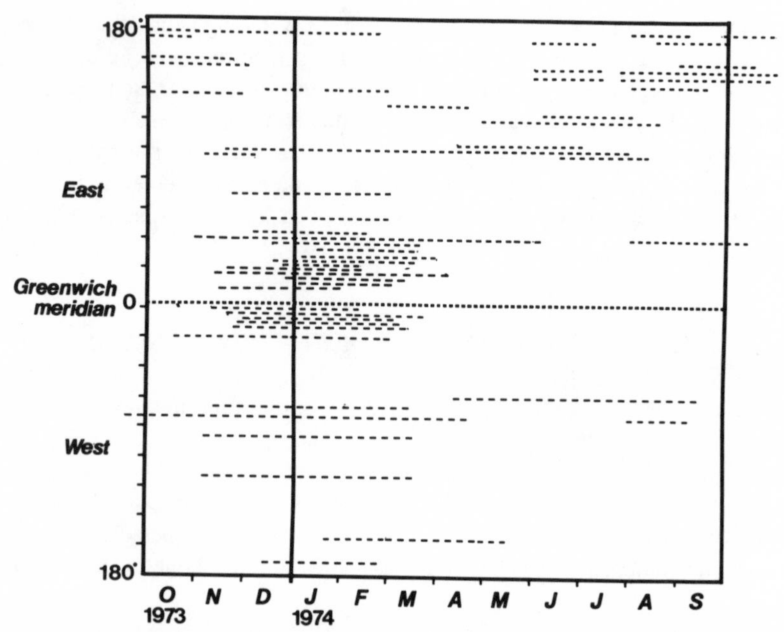

FIGURE 18.5. The epidemics shown in Fig. 18.4 illustrated by longitude instead of latitude. It is difficult to discern any annual pattern such as the transequatorial swing revealed by Fig. 18.4.

ical conundrums posed by epidemic influenza. Moreover, it would be convenient to have all the material aggregated under a single cover together with an historical background and a consideration of alternative hypotheses.

The subject has an even wider interest for students of epidemiology and community medicine because, if the new concept be correct, it carries implications for the epidemiology of all seasonal diseases. The concept originated largely because of the need to explain the seasonal character of influenza epidemics. What was the mechanism that determined their seasonal appearances and disappearances? Influenza had to be recognized as but one among the millions of seasonal crops whose timing is governed by the annual variation in composition, intensity, and duration of solar radiation resulting from the 23.5° tilt of the rotational plane of the Earth in relation to its plane of circumsolar orbit. Among innumerable examples one may mention the familiar annual effect of this influence on weather, climate, the harvest of plants, and the breeding cycles of many animals.

The mechanisms vary whereby the prime cause mediates the seasonal influence and it is important to discover them for each of the seasonal diseases. In Chapter 8 we pointed out that the mechanisms governing the seasonal behavior of the large group of human infections grouped as the common cold must differ from those that determine the seasonal timing of influenza epidemics. When the disease is caused by a pathogen that is transmitted by the bite of an arthropod, the mediating mechanism of the seasonal nature of the disease is clearly the seasonal life history of the vector, but in many instances it may not be clear how the breeding cycle itself is mediated.

The weekly *Communicable Disease Report* and similar periodicals will have made readers familiar with the seasonal variations that occur in the numbers of the isolations of numerous human pathogens. The intermediate operative mechanisms have been identified for relatively few of them, but such information would be valuable in every case, not only in elucidating the epidemiology but also in planning prophylaxis and treatment.

Most of the work on such problems in many parts of the world is concerned with the seasonal behavior of plants and nonhuman animals and is unconnected with medical studies. Few of the papers deal with the seasonal nature of human diseases that are not transmitted by arthropod vectors, so that few are published in medical journals.

Much that has been written in this book about the significance of season in the study of influenza applies equally to the study of all other seasonal diseases, and it might be rewarding if persons, who are in a position to do so, will look more closely at the operative mechanisms that are causing such seasonal behavior.

Cook *et al.*[13] have recently drawn attention to similar difficulties in explaining the seasonal behavior of rotavirus infections of children in a global study of diarrheal diseases.

REFERENCES

1. Kilbourne ED: *Influenza.* New York, Plenum, 1987, pp 268–269.
2. Schulman JL, Kilbourne ED: Seasonal variations in the transmission of influenza, in *Biometerology II.* New York, Pergamon, 1967, pp 83–87.
3. Cliff AD, Haggett P, Ord JK: *Spatial Aspects of Influenza Epidemics.* London, Pion, 1986.
4. Cliff AD, Haggett P, Ord JK *et al: Spatial Diffusion.* Cambridge, Cambridge University Press, 1981.
5. Bartlett MS: Measles periodicity and community size. *JR Statis Soc* 120:48–70, 1957.
6. Black FL: Measles endemicity in insular populations: Critical community size and its evolutionary implications. *J Theor Biol* 11:207–211, 1966.
7. Hope-Simpson RE: Infectiousness of communicable disease in the household (measles, chickenpox and mumps). *Lancet* 2:549–564, 1952.
8. Hope-Simpson RE: The period of transmission in certain epidemic diseases: An observational method for its discovery. *Lancet* 2:755–769, 1948.
9. Angulo JJ, Pederneiras CAA, Sakuma ME *et al:* Contour mapping of the temporal-spatial progression of a contagious disease. *Bull Soc Pathol Exot Fillales* 72:374–385, 1979.
10. Davenport FM: Reflections on the epidemiology of myxovirus infections. *Med Microbiol Immunol* 164:69–76, 1977.

11. Hope-Simpson RE: Influenza 1951. Discussion. *Proc R Soc Med* 44:798–800, 1951.
12. Glezen WP, Six HR, Perrotta DM *et al:* Epidemics and their causative viruses, in Stuart-Harris CH, Potter CW (eds): *The Molecular Virology and Epidemiology of Influenza.* London, Academic Press, 1984, p 19.
13. Cook SM, Glass RI, LeBaron CW *et al:* Global seasonality of rotavirus infections. *Bull WHO* 68:171–177, 1990.
14. Hope-Simpson RE: Simple lessons from research in general practice. Part 7. *PHLS Microbiol Dig* 7(3):74–79, 1990.

Author Index

For some early writers, the location and date(s) of the influenza epidemic are given in parentheses beside the author's name, e.g., (Alnwick, 1782).

Subject Index